Explanation of Frontispiece
Photograph of a portion of the Great Wall of China where it crosses the Niuxinshan
gold mining district (see text Fig. 3.1). The view is taken from atop the Niuxinshan
granite looking west. The prominent cliff in the middle ground is formed by Late
Proterozoic quartzites of the Changcheng (meaning Great Wall) System which over-
lie Archean metabasites along a thrust fault at the base of the cliff (fault F3 in text
Fig. 3.2). Gold is being mined from quartz veins in the Archean rocks on the northeastern side of the Niuxinshan granite, about 1000 m from the site of this photograph.

R.B. Trumbull G. Morteani
Z.L. Li H.S. Bai

Gold Metallogeny
in the Sino-Korean Platform

Examples from Hebei Province, NE China

With 49 Figures and 21 Tables

Springer-Verlag
Berlin Heidelberg New York
London Paris Tokyo
Hong Kong Barcelona
Budapest

Dr. Robert B. Trumbull
Prof. Dr. Giulio Morteani
Lehrstuhl für Angewandte Mineralogie und Geochemie
Technische Universität München
Lichtenbergstraße 4
W-8046 Garching, FRG

Li Zhiliang
Bai Hongsheng
The First Geological Exploration Bureau
Ministry of Metallurgical Industry
Yanjiao, Eastern Beijing
People's Republic of China

ISBN-13:978-3-642-77352-5   e-ISBN-13:978-3-642-77350-1
DOI: 10.1007/978-3-642-77350-1

Library of Congress Cataloging-in-Publication Data
Gold metallogeny in the Sino-Korean platform: examples from Hebei
Province, NE China/R.B. Trumbull . . . [et al.]. p. cm. Includes biblio-
graphical references and index.
ISBN-13:978-3-642-77352-5 (U.S.)
1. Gold ores-China-Hebei Province. 2. Metallogeny-China-Hebei Pro-
vince. I. Trumbull, R.B. (Robert B.)
QE390.2.G65G655 1992 553.4′1′0951152-dc20 92-17971

© Springer-Verlag Berlin Heidelberg 1992
Softcover reprint of the hardcover 1st edition 1992

The use of general descriptive names, registered names, trademarks, etc.
in this publication does not imply, even in the absence of a specific statement, that such names are exempt from the relevant protective laws and
regulations and therefore free for general use.

Typesetting: Best-set, Hong Kong
32/3130-5 4 3 2 1 0 – Printed on acid-free paper

Preface

This monograph reports the results of a cooperative research project conducted from 1987 to 1989 on selected gold deposits in eastern Hebei province, People's Republic of China. The European partners were the Technical University of Munich, Federal Republic of Germany, and the University of Trento, Italy. The Chinese partner was the Ministry of Metallurgical Industry (MMI) in Beijing, which is responsible for the development and operation of all major gold mines in China. In addition to data and observations collected during the project investigations, this monograph includes a great deal of information taken from published and unpublished geologic reports and maps from sources in China.

This research has been realized with financial assistance from the Commission of the European Communities under the EC International Scientific Cooperation Programme with the People's Republic of China. The views herein expressed are those of the participating scientists and do not represent any official view of the EC or of the government of the People's Republic of China.

Questions of Style

We have tried to develop a consistent style in this text while combining information and ideas from both western and Chinese sources. For the writing of Chinese names and localities we use the Pinyin system of transliteration and the following conventions:

a) All Chinese place names are written as one word; thus we write "Niuxinshan" and not "Niu Xin Shan". In the case of personal names, the family name is written first followed by the given name, and the latter is written as one word (e.g., Sun Dazhong).
b) Citations of both Chinese and Western authors are made in the same conventional style. For the Chinese authors,

only the family names are written in full, and given names are represented by one or two initials depending on the number of syllables in the name. For example, Sun Dazhong is cited as Sun D Z, and Li Ren is cited as Li R.

A troublesome problem of style concerns the nomenclature of the Early Precambrian rock units and the terms used to describe the regional tectonics. The Chinese geologic terminology does not conform with modern international practice in naming high-grade metamorphic terranes and it does not reflect modern concepts of plate tectonics. The authors of this book acknowledge this problem but they have neither the inclination nor the competence to update the nomenclature of Chinese regional geology. We therefore follow the nomenclature given in the recent comprehensive references to the geology of China (Yang et al. 1986; Ren et al. 1987). This may irritate some readers, as it has irritated one reviewer of the text, but it has the important advantage that one can easily look up details about specific lithologic or tectonic units cited here in the Chinese literature without having to guess at the original form of their names.

Acknowledgments

The work reported in this text would have been impossible without the financial support of both the European Community and the Chinese Ministry of Metallurgical Industry. The authors gratefully acknowledge this support. The European contribution to this project owes much to the work of Dr. Gerhard Lehrberger, Technical University of Munich, who helped with all field work in China and with a large part of the petrographic and geochemical studies in Munich. Prof. Muharrem Satir, now head of the Institute for Geochemistry, University of Tübingen, directed the isotopic studies of the European partners, and aided in field sampling for age determinations. His successor at the Technical University of Munich, Dr. Dominique Blamart, performed many of the stable isotopic analyses reported here, and made useful comments on this manuscript. We thank Prof. Andreas Fuganti and Dr. Guido Ceri of Geoexpert International, Trento, Italy for their work on LANDSAT satellite interpretations. Dr. Dietrich Ackermand, University of Kiel, gave access to his microprobe laboratory and helped with the microprobe analyses. Finally, the European partners wish to personally thank the staff members of the Ministry of Metallurgical Industry (MMI) for the warm hospitality and efficient organization which made our stays in China a pleasure. In particular we wish to thank Mr. Yao Peihui, Mrs. Hu Pinmei, and Mr. Lu Jin of the Geological Bureau of MMI in Beijing; and Mr. Li Maocai, director of the First Geological Prospecting Bureau of MMI in Yanjiao, for their untiring support of our joint project. A final word of thanks is due Mr. Liu Hua of the First Geological Prospecting Bureau of MMI for his excellent service as translator during our stays in China.

The authors from the Ministry of Metallurgical Industry wish to thank the Analysis Center, Chinese Academy of Geosciences of the Ministry of Geology and Mineral Resources, and the Analysis Center of the Geoscience Institute of Academica Sinica for support in providing chemical analyses and interpretations. For their help in field investigations at the various deposits we thank Mr. Ma Wenrong, engineer of the No. 515 Geological Prospecting Team, First Geological Prospecting Bureau of MMI, for support of investigations in the Niuxinshan district; and Mr. Yuan Yingliang, engineer of the No. 522 Geological Prospecting Team, First Geological Prospecting Bureau of MMI, for guidance in the Sanjia district. For his excellent petrographic work on rock and ore sections we thank Mr. Yang Jun, engineer of the Research Institute for Geology and Exploration, First Geological Prospecting Bureau of MMI. We also are indebted to Mr. Zhang Haowen, assistant engineer of the same unit, for help with sample preparation, drafting, and data processing.

The manuscript has benefited greatly from reviews by Prof. Brian Windley, Prof. Ian Plimer, and Prof. Francis Saupé, all of whom are heartily thanked for their comments.

Summer 1992   R.B. Trumbull
                                           G. Morteani
                                           Z.L. Li
                                           H.S. Bai

Contents

1 Introduction

Many of the world's most important gold fields occur in Archean and Early Proterozoic host rocks. The most common association of gold in this setting is with the metamorphosed supracrustal volcano-sedimentary series of the greenstone belts. Prominent examples of the greenstone-gold association are known from nearly all of the Archean cratons, including those in South Africa (Barberton Belt), Zimbabwe (Sebakwian and Bulawayan Belts), Ghana (Birrimian Belt), Western Australia (Norseman-Wiluna Belt), Canada (Abitibi Belt), India (Kolar Schist Belt), and Brazil (Sao Francisco Craton). Reviews of gold geology by Foster (1991), Keays et al. (1989), Hutchinson and Vokes (1987), Foster (1984), and Boyle (1979) provide extensive documentation of this type of gold deposit in the above-mentioned areas.

Conspicuously missing in the international literature are descriptions of the important gold deposits in the Archean craton of northeastern China (with a few exceptions, such as Sang and Ho 1987). According to sources cited by Sang and Ho (1987), China was the world's fourth largest gold producer, with over 3000 deposits. The U.S. Bureau of Mines' estimate of gold production in China in 1989 was 80 tons (U.S. Bureau of Mines 1990). The agency responsible for the operation of gold mines in China depends on their size. The largest mines are operated at the state government level by the Ministry of Metallurgical Industry. Medium-sized and small mines are operated by units of the local county governments.

According to Sang and Ho (1987) and Zhu (1989), the majority of gold production in China comes from deposits in Precambrian uplifts of northeastern China (Shandong, Hebei, Liaoning, and Jilin provinces). The Archean and Early Proterozoic supracrustal rocks in this area, and the gold deposits they contain, are similar in many respects to those of the Archean greenstone belts in Canada, Western Australia, and southern Africa. However, the past decade of research, mostly by Chinese scientists, has shown that there are important differences between the Archean-hosted gold deposits in China and the greenstone belt gold deposits found in other continents.

The need for a metallogenetic model specific to the deposits in northeastern China was recognized by the Ministry of Metallurgical Industry, and this was the main goal of the research on which this text is based. Our investigations were centered around the following six questions considered to be critical for understanding the genesis of these deposits:

1

1. The age of mineralization.
2. The source of the gold.
3. The role of metamorphic iron formations is gold metallogenesis.
4. The role of granitic intrusions in gold metallogenesis.
5. The structural control of the deposits.
6. The physicochemical conditions of mineralization and the nature of the ore-forming fluids.

This book begins with a brief description of the gold metallogenetic provinces in northeastern China and an introduction to the gold districts of Eastern Hebei province. Chapter 2 summarizes the regional geology of eastern Hebei province with emphasis on the tectonic setting and on the nature of the Precambrian basement. The third chapter then discusses the geological, mineralogical and geochemical nature of five gold districts which were chosen to represent the range of Archean-hosted gold deposit types in this region. Chapter 4 discusses the controls of gold metallogeny based on these examples, and Chapter 5 formulates a tentative metallogenetic model for the deposits. In the final chapter, the Chinese gold deposits are compared with the Archean greenstone-gold deposits of other continents.

1.1 Regional Tectonic Division of China

Before discussing the major gold mining provinces of northeastern China, it is useful to introduce some aspects of the regional geology. The tectonic division of China by Huang (1945) distinguished between stable "platforms" and intervening and/or marginal "mobile" (or fold) belts, and his main tectonic units are, with some modification, still used today. The most recent tectonic compilations are the *Tectonic Map of China* at a scale of 1:4 000 000 (Chinese Academy of Geological Sciences 1979) and the books by Yang et al. (1986), Ren et al. (1987), and Chen (1989). The *Tectonic Map of Asia* at a scale of 1:8 000 000 (Li et al. 1982) combines the tectonic features of China with the rest of Eurasia east of the Mediterranean, and presents a plate tectonic interpretation. Figure 1.1 shows a simplified map of the main tectonic units of China based on Ren et al. (1987), where the platforms and "fold systems" of various ages are clearly shown. Each "fold system" on the figure is made up of one or more "fold belts", and the reader is referred to Ren et al. (1987) and references therein for more complete information.

The tectonic unit which contains the gold districts of eastern Hebei province is the Sino-Korean Platform and the following sections on orogenies, structure, and magmatism focus on this unit. As shown in the following section, other important gold provinces in northeastern China occur in basement uplifts within the marginal fold belts north of the Sino-Korean Platform (e.g., subunits of the Inner Mongolian-Great Hinggan and Jilin-Heilongjiang fold systems in Fig. 1.1).

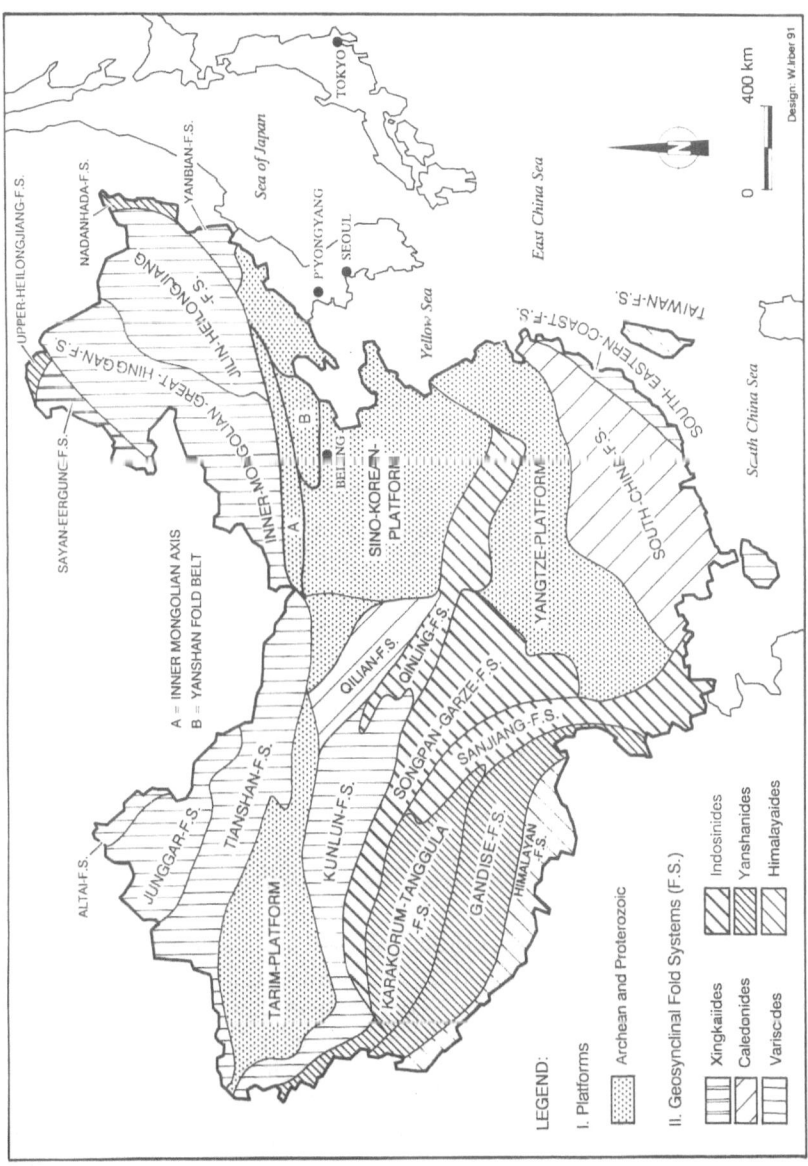

Fig. 1.1. Tectonic map of China. (After Ren et al. 1987)

The Sino-Korean Platform is characterized by several massifs of Early Precambrian metamorphic basement rocks which range from Archean to Late Proterozoic in age, and continental or marine sedimentary cover rocks of Late Proterozoic to Cenozoic age. The Early Precambrian basement massifs (see Fig. 2.1) may represent individual continental "nuclei" which were brought together by later orogenies (Yang et al. 1986). The major part

3

of the platform was consolidated before the end of the Middle Proterozoic, as shown by the extent of Late Proterozoic and younger sedimentary rocks of platformal type (Ren et al. 1987).

The Sino-Korean Platform has not been completely stable since its consolidation in the Proterozoic, and for this reason the term "paraplatform" is preferred by many authors (e.g., Yang et al. 1986; Ren et al. 1987). The influence of Paleozoic (Caledonian and Variscan), Mesozoic (Indosinian and Yanshanian), and Cenozoic (Pacific margin) orogenies is strong in the marginal parts of the platform (Ren et al. 1987). Tectonic reactivation of the platform takes the form of so-called platformal fold belts, deep-seated fracture systems, and zones of magmatism. Two such reactivated units, the so-called Inner Mongolian Axis and Yanshan Fold Belt are singled out in Fig. 1.1A and B, respectively because they are the hosts of the gold deposits discussed in this text.

The strongest orogenic disturbance of the Sino-Korean Platform occurred in the Late Mesozoic (Yanshanian Orogeny), when extensive magmatic activity and NE-trending structures formed as a result of plate interactions on the Pacific margin (Takahashi 1983). It will be shown below that this activity was important for gold metallogenesis in eastern Hebei province and adjacent areas. Cenozoic tectonism on the eastern margin of the Sino-Korean platform was characterized by extension, subsidence, and basin development with associated fissure eruption of alkaline basalts. At this time, the marginal seas (Bohai, Yellow Sea, Sea of Japan) formed by subsidence and extension (Li et al. 1982; Ren et al. 1987). The eastern margin of the platform is still seismically active, and some of the most catastrophic earthquakes in human history have taken place in northeastern China (e.g., Tangshan in 1976).

An important emphasis of recent work by Chinese and foreign geologists has been the application of plate tectonics concepts to interpret the tectonic history of China (Uyeda and Miyashiro 1974; Dickinson 1979; Li et al. 1980, 1982; McElhinny et al. 1981; Klimetz 1983; Terman 1984; Zhang et al. 1984; Maruyama et al. 1989; Wiley et al. 1990). Most studies have focused on the development of the Himalayan orogen (references in Windley 1984) and on the development of the Pacific margin tectonic belt, including the offshore islands and island-arc systems (Uyeda and Miyashiro 1974; Dickinson 1979; Takahashi 1983; Maruyama et al. 1989; Wiley et al. 1990). In older rocks and in regions of interior China far from the continental margins, paleomagnetic evidence and the geologic recognition of paleo-sutures of various ages provide important evidence of plate motions (Zhang et al. 1984). The interpretation of the tectonics of China is still evolving and, although tentative models of the Phanerozoic assembly of eastern Asia have been made (see Li et al. 1982; Klimetz 1983; Zhang et al. 1984; Maruyama et al. 1989; Wiley et al. 1990), these are still highly speculative and not dealt with here in detail. Some aspects of plate tectonics in relation to the Yanshanian Orogeny (Jurassic-Cretaceous) are discussed in Chapter 2.4.

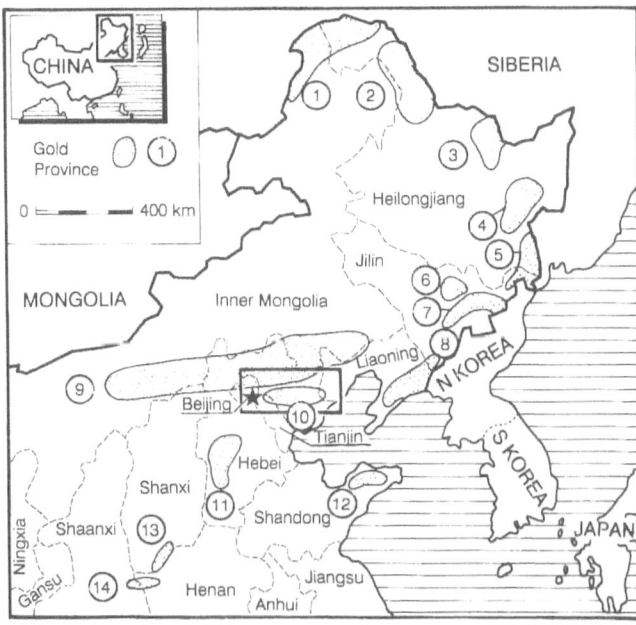

Fig. 1.2. Map of northeastern China showing the distribution of gold provinces discussed in the text. The area of eastern Hebei province covered in this text is *boxed*

1.2 The Gold Provinces of Northeastern China

Important sources of information on the mineral deposits of China in general are Ikonnikov (1975) and the 1:4 000 000 *Metallogenic Map of Endogenic Ore Deposits of China* (Guo 1987). A brief description of some of the most important gold deposits in northeastern China was given by Sang and Ho (1987). Other important sources of information on Chinese gold deposits are the proceedings volumes of the International Conference on Gold Geology and Exploration in Shenyang (Guan and Zhu 1989), and a recent series of monographs entitled *Contributions to the Project of Regional Metallogenetic Conditions of Main Gold Deposit Types in China* edited by the Shenyang Institute of Geology and Mineral Resources (Shenyang 1988a,b, 1989a–d).

Figure 1.2 shows the distribution of 14 gold metallogenetic provinces in northeastern China which were distinguished by Zhang (1979) based on the distribution (i.e., clustering) of deposits, regional tectonic features, geologic environment, and mineralization types. A brief summary of the geologic setting and the mineralization types present in each province is given below. The numbers given in parentheses correspond to the numbers in Fig. 1.2. Where no reference is given, the information cited was obtained from

5

unpublished reports and personal communications of geologists from the Ministry of Metallurgical Industry.

Sang and Ho (1987) discussed five gold metallogenetic provinces in northeastern China. The correlation of these five provinces with the ones delineated below is as follows:

Zhaoye (Sang and Ho) = Ludong (12)
Jiapigou (Sang and Ho) = Longgang-Mudanling (7)
Eastern Hebei (Sang and Ho) = Yinshan-Nuiu'erhushan *and* Yanshan
                              (9 and 10)
Little Qin Hill (Sang and Ho) = Xiaoqinling (14)
Southern Liaoning (Sang and Ho) = Yingkou-Kuandian (8).

E'erguna Gold Province (1)

The E'erguna gold province is located at the northern border of China in an area of Caledonian and Variscan fold belts (the Heilongjiang and E'erguna fold belts). A metamorphic basement contains rocks of mainly Early Cambrian to Middle- and Late Proterozoic age. Variscan granitic intrusions cover more than half of the exposed area. Jurassic volcanic rocks are found in graben structures with NE–SW and E–W strike. Granite of Yanshanian age (Late Mesozoic) intruded in minor amounts along NE–SW-trending regional fault zones. Auriferous quartz-sulfide veins are found at the intersections of NW–SE and NE–SW-trending faults and around the Yanshanian granitic intrusions. Paleoplacer deposits, mainly in Jurassic sandy conglomerates, are also important gold sources, and more than ten deposits of placer gold are known. Further information on the deposits in this province can be found in Shenyang (1988a).

Xiaoxing-Anling Gold Province (2)

The Xiaoxing-Anling gold province lies in the northern part of the Variscan Daxing-Anling fold belt. The basement rocks consist of Proterozoic schists and gneisses. These are covered by Paleozoic phyllites, immature sandstones, shales, and quartzites, and by Mesozoic andesite-rhyolites. A large Caledonian granite intrusion occurs in the northwestern part of the province, whereas Variscan granites dominate in the eastern and northern parts. Smaller granite intrusions of Yanshanian age are scattered throughout the province. NNE- and NE-trending fault zones influence the distribution of both the Variscan and Yanshanian granite intrusions and of the volcanic rocks. Gold-bearing quartz-sulfide veins constitute the most prominent type of deposit in the Precambrian host rocks, but in the younger volcanic and plutonic rocks epithermal mineralizations, skarn-type deposits, and porphyry-type deposits are also present. Most of the primary gold deposits are associated with the contact zones of the granites; however, the main gold production from the province is from placer gold deposits, with about ten producing mines.

6

Jiayin-Luobei Gold Province (3)

The Jiayin-Luobei province is located in the northern part of the Jiamosi uplift in a Variscan fold belt north of the Sino-Korean Platform. In the western part of the province widespread Variscan granites and minor occurrences of Proterozoic and Paleozoic sedimentary rocks are found in the Qingheshan Uplift. The central part of the province is formed by a Mesozoic fault-bounded basin (Wulaga Basin). In the eastern part of the province, large areas of Proterozoic schists and gneisses are exposed in the Taipingguo Anticline. Mesozoic rocks occur on both flanks of the anticline. The dominant intrusive rocks are Yanshanian granites, although granites of Variscan age are common in the western part. During the Late Cretaceous, dacites and rhyodacites intruded at a subvolcanic level. These subvolcanic rocks locally host a porphyry-type gold mineralization. Both the plutonic bodies and the subvolcanic intrusions are concentrated along the faults which border the Wulaga Basin. The oldest faults strike NW–SE, and were active intermittently over a long time. E–W- and N–S-trending oblique reverse faults control the distribution of the Mesozoic volcanic rocks. NE–SW- and E–W-trending oblique reverse faults gave passage to the Yanshanian granitic intrusions.

The main gold deposits in the Jiayin-Luobei province are porphyry-type deposits, stratiform metamorphic deposits, and Quaternary placer deposits. The first type is represented by the Tuanjiegou deposit (Li and Liu 1986). The Dongfengshan deposit is a typical stratiform metamorphic gold deposit (Liu 1987; Hao 1989). Other deposits in this province are described in Zhu (1979) and Shenyang (1988a).

Huanan-Laoyeling Gold Province (4)

The Huanan-Laoyeling province includes a fault-bounded Mesozoic sedi-mentary basin (Wokenhe Basin) in the southern part and Proterozoic basement rocks in the northern part. The latter consist of anatectic granites and metamorphic rocks of Early Proterozoic age exposed in a complex NW–SE-trending anticlinorium. The Wokenhe Basin in the southern part of the province contains a Jurassic carbonaceous clastic sequence and Cretaceous volcanic rocks. The main magmatic rocks are migmatitic granite and granitic gneisses of Early Proterozoic age and Yanshanian granitic intrusions. The regional faults in the Huanan-Laoyeling gold province strike NW–SE and NE–SW. These faults are important, since they form the border of the Mesozoic basin and control in part the distribution of Yanshanian granites and mineralization zones.

Gold mineralization occurs as hydrothermal quartz-sulfide veins in and around the Yanshanian granitic plutons and in the Early Jurassic sandstone-conglomerate series. Placer deposits are also important, and in fact the best-known gold deposit in the province is the placer deposit of the Huanan River. Further information on the deposits in this province can be found in Zhu (1979) and Shenyang (1988a).

Taipingling Gold Province (5)

The Taipingling province is situated in the Yanbian fold belt of Variscan age. In the northern section, schists and gneisses are exposed but the province in general consists of Carboniferous and Permian clastic and volcanic rocks and of Mesozoic volcanics. The most common intrusive rocks are of Variscan age, and they cover a wide area. Granites of Early Proterozoic age occur in the northern section, and Yanshanian granites form only minor bodies which extend along NE–SW-trending fractures.

The most common type of gold deposit in this province consists of hydrothermal quartz-sulfide veins centered about small Variscan granite intrusions. Also mined are epithermal gold-silver deposits hosted by volcanic rocks, and Quaternary placer gold deposits. The primary gold deposits are clearly associated with fracture zones. The gold deposits in the northern section are spatially related to the Variscan granite bodies, and they occur along NE–SW- and NNE–SSW-trending fracture zones. In the central part of the province, gold mineralization is concentrated in N–S-trending fracture zones and is mainly hosted by Permian volcanic rocks and Carboniferous sedimentary rocks. In the southwestern part of the province, gold mineralization is located near the marginal faults of Mesozoic basins. These faults trend mainly E–W and NW–SE. Further information on the deposits in this province can be found in Zhu (1979) and Shenyang (1988a).

Hadaling Gold Province (6)

This province is situated along the northern part of the Juifa River at the northern border of the Sino-Korean Platform in a Variscan fold belt. The geologic units consist of Paleozoic sedimentary and minor volcanic rocks which were folded and overturned at the end of the Permian. Magmatism is represented by granitic intrusions of Variscan and Yanshanian age. The intrusions occur along NE–SW- and NW–SE-striking faults, and they influence the distribution of gold deposits.

Two main areas or subprovinces of gold mineralization occur: in the Erdaodianzi subprovince the dominant ore-controlling structure is an anticlinal fold belt with mainly E–W-trending axes. Gold occurs in quartz saddle-reef veins in the fold hinge zones. The most common host rocks of the gold-bearing veins are Permian carbonaceous shales and hornfelses in the vicinity of a Variscan granite intrusion. Placer gold deposits of Quaternary age are of some importance in this area. The Yongji-Panshi subprovince is built up by Late Carboniferous metavolcanic rocks and Permian carbonaceous sedimentary rocks. Jurassic sedimentary rocks are found in basins bordered by NE–SW- and NW–SE-trending faults. Gold occurs in quartz veins within the Carboniferous metavolcanics. Tang (1986) gives further information on the deposits in this province.

Longgang-Mudanling Gold Province (7)
This gold province is located at the northern edge of the Sino-Korean Platform in the northern part of the Jiaoliao Uplift. The Early Precambrian basement includes amphibolites, plagioclase-hornblende gneisses, greenschists and magnetite-rich quartzites (metamorphic iron formation). Carboniferous and Permian sandstones and shales and Mesozoic volcanic rocks occur locally. Large granite bodies of Variscan age intruded along NW–SE faults mainly in the northeastern part of the province but they are also found as smaller stocks throughout the province. Yanshanian granite bodies are irregularly distributed throughout the province.

Gold mineralization is mainly associated with NE–SW- and NW–SE-trending regional faults. Gold-quartz veins in the Precambrian metamorphic rocks are the dominant type of mineralization. A typical example of this type is the Jiapigou gold deposit (Hu 1989; Nesbitt 1991). Quaternary placer gold deposits are also mined. More information on these deposits is given by S.Q. Wu (1985) and Sang and Ho (1987).

Yingkou-Kuandian Gold Province (8)
The Yingkou-Kuandian gold province is situated in the Jiaoliao Uplift in southern Liaoning province. Within the Jiaoliao Uplift, a Precambrian basement is exposed which consists mainly of high-grade metamorphic rocks of the Archean Anshan Group and Early Proterozoic greenschist-facies metasedimentary rocks of the Liaohe Group. Granite bodies of Middle Proterozoic age cover wide areas. Yanshanian granitic intrusions are also common in a zone of NE–SW elongation.

The gold mineralization is best developed in the Archean and Proterozoic metamorphic rocks. The main types of deposits are gold-quartz veins, pervasively altered fracture zones, metamorphic stratiform deposits and skarn-type mineralizations at the contacts with Yanshanian granites. The gold deposits are mainly associated with E–W-, NNE–SSW-, and NE–SW-trending structures and related secondary fracture and shear zones. Intersections of major fracture zones are particularly important. Placer gold deposits of Quaternary age are of minor economic importance. For more information on the deposits in this province see Sang and Ho (1987) and Shenyang (1988b).

Yinshan-Nuiu'erhushan Gold Province (9)
The Yinshan-Nuiu'erhushan province is a large gold province which extends along the northern border of the Sino-Korean Platform. Most of the province is located within a zone of uplifts known as the Inner Mongolian Axis (Fig. 1.1A) in which Early Precambrian basement is exposed. The basement includes Late Archean rocks which consist of plagioclase-hornblende gneisses, amphibolites, gneisses, schists, marbles, and metamorphic iron formation. Early Proterozoic metamorphic rocks crop out in the western part of the province, and Middle- to Late Proterozoic platformal sedimen-

tary rocks are found in isolated occurrences in the eastern part. Neritic and littoral marine, and intercalated terrestrial sedimentary rocks of Paleozoic age occur on both flanks of the uplift. Mesozoic basins bordered by deep-seated faults contain coal-bearing sedimentary sequences with important volcanic components. Igneous rocks of several ages are present. Middle- and Late Proterozoic intrusions and volcanic rocks span the compositional range from ultrabasic through acid to alkalic. The Proterozoic intrusions are found mainly in the western part of the province. Late Paleozoic and Mesozoic (Variscan and Yanshanian ages), intermediate to acid intrusives are widely distributed. Yanshanian intrusive and extrusive rocks are particularly abundant in the central and eastern parts, and the Yanshanian intrusions are closely related to gold mineralizations.

The main types of gold deposits in the Yinshan-Nuiu'erhushan province are gold-quartz veins and mineralized shear zones in the Archean metamorphic rocks. A spatial association of the deposits with Yanshanian granite intrusions is common. Some vein-type deposits are found within, or in the contact zones of Variscan, and especially, Yanshanian intrusions or within Jurassic volcanic rocks. In addition, mesothermal and epithermal volcanic-related deposits, skarn-type deposits, and Quaternary placer gold deposits are known. The mineralized structures strike E–W, NE–SW, and NNE–SSW. Intersection zones are of major importance for mineralization. Ore bodies are mainly found in secondary fractures related to the regional faults. Important primary gold deposits in the Yinshan-Nuiu'erhushan province are the Jinchanggouliang, Honghuagou, Dashuiqing, and Xiaoyingpan deposits. One of the famous placer gold deposits is the Jinpen deposit, situated in a Mesozoic-Cenozoic basin in the western part of the province. Further information on the deposits in this province can be found in Xia (1986), Guan et al. (1989), and Trumbull et al. (1990).

Yanshan Gold Province (10)

The Yanshan gold province includes the deposits of eastern Hebei province which are the main subject of this book. The gold province is located within the Sino-Korean Platform in the so-called Yanshan platformal fold belt (Fig. 1.1B), which exposes an Early Precambrian basement. The geologic units are dominated by high-grade metamorphic rocks of Archean and Early Proterozoic age. These Early Precambrian rocks are exposed in anticlinoria surrounded by Middle- and Late Proterozoic low-grade metasedimentary rocks of platformal type. Jurassic rocks are distributed in rare outcrops in the northeastern part of the province and they represent a suite of continental and volcanoclastic basin sediments. The earliest magmatic activity recognizable in the Yanshan gold province is represented by Precambrian ultramafic, mafic, and intermediate intrusions associated with E–W-trending structures, mainly in the eastern and western parts of the province. Many of these rocks have been metamorphosed, but textures and field relations clearly reveal their igneous character. A second phase of magmatism took place during the Mesozoic, and the peak magmatic activity was during the Yanshanian Orogeny. Intrusive rocks of this age are mostly of granitic composition. They form stocks of various sizes, often occurring along fracture zones trending NE–SW and NNE–SSW. Igneous dikes of felsic to mafic composition also formed during the Late Yanshanian Orogeny.

Gold mineralization occurs in gold-quartz veins and pervasively altered fracture zones which are controlled by secondary fractures related to regional-scale NE–SW- and NNE–SSW-trending fault systems. Most gold deposits are spatially associated with granitic intrusions and dikes of Yanshanian age. The most common host rocks are Archean and Early Proterozoic metamorphic rocks. Less important host rocks are the Yanshanian granites. The largest primary gold deposits in the Yanshan province are the Jinchangyu and Yuerya deposits, which are discussed in detail in Chapter 3 of this book. Quaternary placer gold deposits are also worked locally. Further information on the deposits in this province can be found in Sang and Ho (1987), Li (1988), and Shenyang (1989a).

Wutai-Taihang Gold Province (11)

The Wutai-Taihang gold province is located in the interior of the Sino-Korean Platform on the east side of the Shanxi Uplift. The basement rocks include Early Archean high-grade metamorphic sequences overlain unconformably by Middle- and Late Proterozoic low-grade metasedimentary rocks. Cambrian and Ordovician strata occur outside the basement uplift. Magmatism is of only minor importance in the Wutai-Taihang province. As in most areas in the eastern part of the Sino-Korean platform, the magmatic rocks belong to two age groups, Middle Proterozoic and Mesozoic (Yanshanian). The Proterozoic granites are known only from scattered occurrences in the Wutai mountains. The Yanshanian rocks are concentrated in the southeastern and northeastern parts of the province.

Gold mineralization is closely related to Yanshanian felsic dikes which occur along NNW–SSE-trending fault zones. The main types of gold mineralization are syn-metamorphic and post-metamorphic hydrothermal gold-quartz veins in Archean host rocks, and Quaternary placer deposits. The most important deposits of the Wutai-Taihang province are the Yixinzhai and the Shihu deposits. Zhang (1986) describes one of the important deposits (Gengzhuang) in this province.

Ludong Gold Province (12)

The Ludong gold province is the most important in northeastern China in terms of gold production and reserves. It is situated in the eastern Sino-Korean Platform within the so-called Jialiao Uplift. The basement rocks in the province are Late Archean and Early Proterozoic high-grade metamorphic rocks. Cretaceous sandstones and shales, conglomerates, and marls occur within extensional basins, and at the basin margins Late Mesozoic and Tertiary felsic volcanic rocks are found. The magmatic rocks in the Ludong province are dominated by granitic intrusions of Yanshanian age. The main structures in the province are E–W-trending folds in the basement complex and large-scale NE–SW- and NNE–SSW-trending fault zones of Mesozoic age. The western part of the gold province is bounded by the major NE–SW-trending Tancheng-Luliang fault.

The gold mineralization is mainly associated with NE–SW-trending Mesozoic faults and related secondary fractures in the granites. The mineralization types are gold-quartz veins and altered fracture zones. The main primary gold deposits in the province are the Linglong and Jiaojia deposits. Some placer gold deposits are also known. Descriptions of the deposits in this province can be found in Huang (1986), Sang and Ho (1987), Shenyang (1989c), Lu et al. (1989), and Zhou and Fan (1989).

Zhongtiaoshan Gold Province (13)

The Zhongtiaoshan province is located in the northern part of the western Henan uplift at the southern margin of the Sino-Korean Platform. The basement consists of Late Archean metamorphic rocks including mica schists, amphibolites, and hornblende gneisses; and Early to Middle Proterozoic quartzites, conglomerates, and marbles with minor mica-schists and phyllites. The main structures of the province trend NNE–SSW. The basement structures include tight folds overturned to the west. Faults are mainly NNE-directed overthrusts and E–W-trending fracture zones. Magmatism is dominated by Yanshanian granitic intrusions and associated felsic dikes concentrated in a zone along the northern part of the province.

Gold-quartz veins and auriferous porphyry copper deposits are the main types of gold mineralization. Whereas the quartz veins are mostly found in the basement rocks at least 10 km away from Yanshanian intrusions, the porphyry-type deposits formed within or in the contact zone of granodiorites. Gold placers also occur in the province. Further information can be found in Sha (1986).

Xiaoqinling Gold Province (14)

This province lies in the contact zone between the Sino-Korean Platform to the north and the so-called Qinling fold belt to the south. The latter is a complex zone of superimposed Paleozoic and Mesozoic folds situated between the Sino-Korean and the Yangtze Platforms. Basement rocks are exposed in an E–W-trending anticlinorium bounded by major fault zones. The rocks consist of high-grade Late Archean amphibolites, migmatitic gneisses, and metamorphic iron formations; and Early Proterozoic metavolcanic rocks, overlain by nonmetamorphosed Late Proterozoic clastic and carbonate sedimentary rocks. Jurassic and Tertiary sedimentary rocks occur outside the basement uplift. Magmatic rocks are represented by Proterozoic granites situated at the southern margin of the province and by granite intrusions of Yanshanian age within the anticlinorium and along the bordering faults.

12

Thrust faults are the main mineralized structures and they are mostly developed in the axial zone of a secondary anticlinorium. Less important for the mineralization are N–S-trending oblique normal faults. The gold mineralization is closely related to granites of Yanshanian age. Further information on the deposits in this province can be found in Wang (1987), Sang and Ho (1987), and Shenyang (1989b).

1.3 Gold in Eastern Hebei Province

The gold deposits of eastern Hebei province are included in the so-called Yanshan Gold province described above (number 10 in Fig. 1.2). Yu et al. (1989) reported 197 gold deposits and occurrences in eastern Hebei province. The distribution of the most important of these is shown in Fig. 1.3. Two important features of the distribution of gold deposits are apparent from the map. First, the deposits are mainly found in the Archean and Early Proterozoic basement and second, many deposits show a spatial association with Mesozoic granitic intrusions.

Two main gold deposits types (excluding placers) occur in eastern Hebei province according to the classification of Zhu (1989), namely, metamorphic-hosted and granite-hosted deposits. The metamorphic-hosted type is by far the more common, and most gold deposits occur in Archean mafic metamorphic rocks. Both the metamorphic-hosted and granite-hosted gold deposits are associated with secondary faults and fractures related to major district-wide fault zones and/or lineaments.

The largest gold mining districts in Eastern Hebei province are briefly described below. The letter preceding each district name corresponds to those on Fig. 1.3. For the purpose of location, the nearest village or town is given in the descriptions, although these towns cannot be shown at the scale of Fig. 1.3.

A) Niuxinshan
The Niuxinshan gold mining district is located about 40 km west of the town of Qinglong. The mineralization is developed in quartz-sulfide veins surrounding a Yanshanian granite intrusion. The host rocks are Early Archean amphibolites (Qianxi Group), and to a lesser extent, Yanshanian granite. Two deposits occur in the district, the Niuxinshan and Huajian deposits. A detailed description of the Niuxinshan district is given in Chapter 3.1.

B) Sanjia
The Sanjia gold mining district is located about 20 km north of the town of Qinglong. It has the same geologic character as the Niuxinshan district. Quartz-sulfide veins with gold are found in Early Archean amphibolites (Qianxi Group), and rarely in Yanshanian granite. Three gold deposits

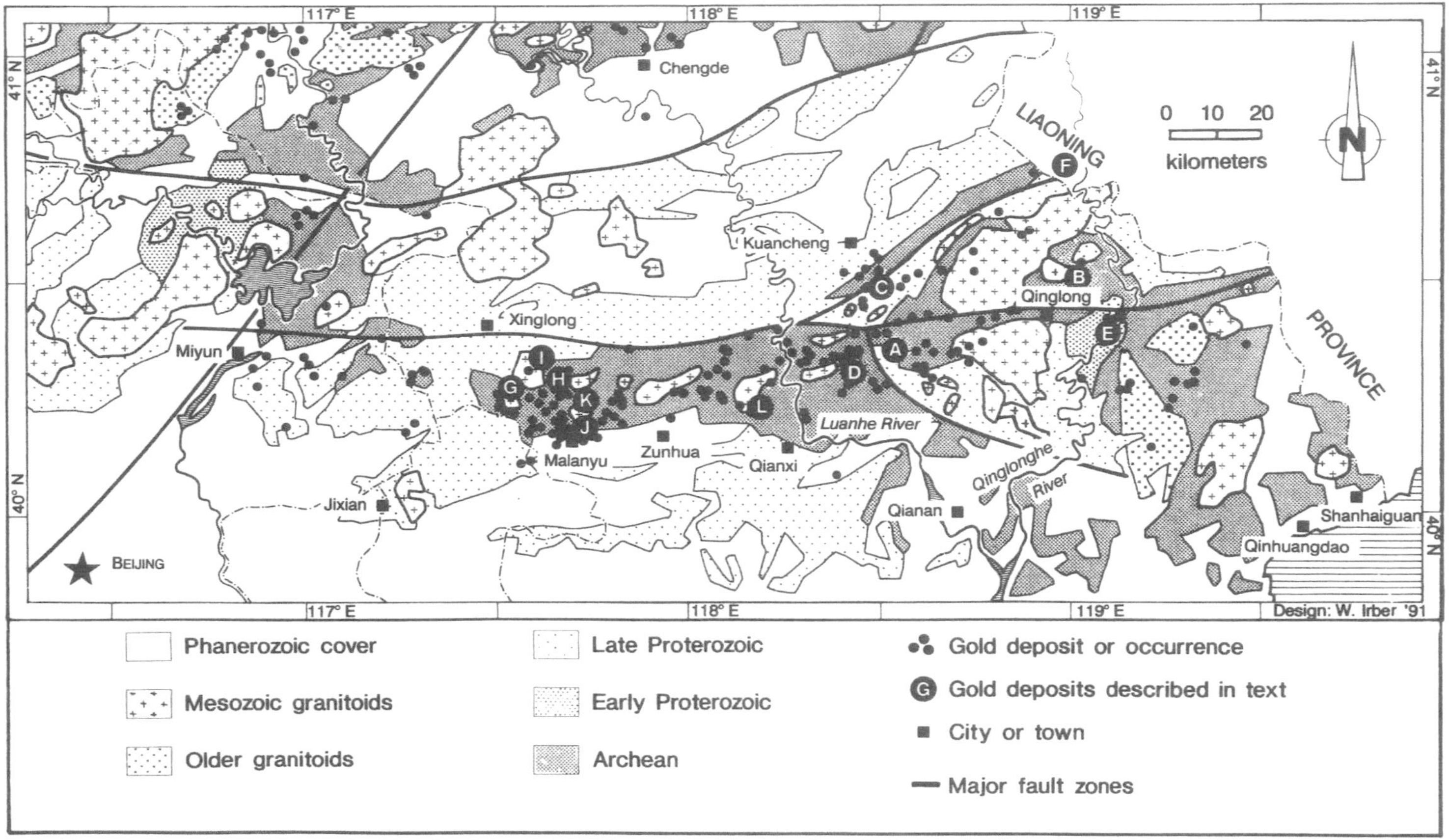

117° E
118° E
119° E
41° N
40° N
Chengde
LIAONING
PROVINCE
0 10 20
kilometers
N
Kuancheng
Qinglong
Xinglong
Miyun
Luanhe River
Qinglonghe River
Zunhua
Qianxi
Malanyu
Jixian
Qianan
Shanhaiguan
Qinhuangdao
BEIJING
A
B
C
D
E
F
G
H
I
J
K
L
Design: W. Irber '91
Phanerozoic cover
Mesozoic granitoids
Older granitoids
Late Proterozoic
Early Proterozoic
Archean
Gold deposit or occurrence
Gold deposits described in text
City or town
Major fault zones

occur within the Sanjia district: Sanjia, Xinglonggou, and Wangtoushan. The deposits are described in detail in Chapter 3.2.

C) Yuerya

The Yuerya gold mining district is located near Yuerya village in Kuancheng county. The mineralization consists of hydrothermal quartz-pyrite veins and vein-parallel disseminated zones near the contact of a Yanshanian granitic intrusion and Late Proterozoic dolomitic limestone (Changcheng system). The main host rock is granite. The Yuerya deposit is described in detail in Chapter 3.3.

D) Jinchangyu

The Jinchangyu district is located southwest of Xiaying village in Qianxi county. The gold mineralization is mainly developed in hydrothermal polymetallic quartz veins in shear zones. The main host rocks are Early Archean amphibolites and plagioclase-hornblende gneisses (Qianxi Group). A detailed description of the district is given in Chapter 3.4.

E) Banbishan

The gold mining district of Banbishan is located near Shuangshanzi village in Qinglong county. The mineralization consists of thin quartz-sulfide stringers and disseminations in shear zones. The host rocks are Early Proterozoic quartz-rich schists and conglomerates (Zhuzhangzi Group). The district includes the Banbishan and Zhangzhangzi deposits, or which only the Banbishan deposit is currently mined. A detailed description of the Banbishan deposit is given in Chapter 3.5.

F) Baizhangzi

The Baizhangzi district is located near Baizhangzi village in Lingyuan county. The mineralization consists of hydrothermal quartz veins related to a Yanshanian granite intrusion. The ore veins are hosted partly by Yanshanian granite and partly by Middle Proterozoic sandstone (Changcheng system).

G) Madi

The Madi district is located in the Paomaochang commune in Xinglong county. The mineralization occurs in quartz veins hosted by granoblastic gneisses and amphibolites of the Archean Qianxi Group.

◀ **Fig. 1.3.** Simplified geologic map of eastern Hebei province with gold deposits and occurrences marked by *black dots*. The deposits described in the text are: *A*, Niuxinshan; *B*, Sanjia; *C*, Yuerya; *D*, Jinchangyu; *E*, Banbishan; *F*, Baizhangzi; *G*, Madi; *H*, Huashi; *I*, Daoliushui; *J*, Malanyu; *K*, Maoshan; *L*, Majiayu

H) Huashi

The Huashi gold mining district is located 4 km west of the Sibazi commune in Xinglong county. The mineralization occurs in quartz veins within faults developed in Early Archean granoblastic gneisses.

I) Daoliushui

The Daoliushui gold mining district is located near Jinshan commune in Xinglong county. Gold mineralization occurs in hydrothermal quartz-sulfide veins hosted by Yanshanian granites and by Archean amphibolites and granulites (Qianxi Group).

J) Malanyu

The Malanyu gold district is located near the village of Malanyu about 20 km northwest of Zunhua. The gold is worked from several Quaternary river terrace sediments. The main gold enrichment is found in sandy and conglomeratic sediments. Underlying the sediments are Archean metamorphic basement rocks and Yanshanian granodioritic intrusions.

K) Maoshan

The Maoshan gold deposit is located near the village of Maoshan in Zunhua county. The gold mineralization occurs in quartz-sulfide veins in Archean metamorphic rocks near the contact of the Yanshanian Maoshan Granite. Some ore veins also occur within the granite.

L) Majiayu

The Majiayu district is located near the Gaojiadian commune in Qianxi county. Gold mineralization occurs in quartz-sulfide veins hosted partly by Early Archean amphibolites and leptites of the Qianxi Group, and partly by Middle Proterozoic quartzites and marbles.

2 Geologic Framework of the Sino-Korean Platform

The purpose of this chapter is to summarize the geologic setting of the gold deposits in eastern Hebei province within the context of the Sino-Korean Platform. Following a brief summary of the main orogenic events which affected the Sino-Korean Platform, this discussion concentrates on aspects of the geology which are most relevant to the genesis of the gold deposits, namely:

1. the lithology and metamorphic history of the Precambrian basement,
2. the structural geology of the Precambrian basement and the timing and orientation of major fault zones,
3. the Yanshanian (late Mesozoic) magmatism, which in this area involved dominantly granitic intrusions and dikes.

A complete summary of the geology of the Sino-Korean Platform is beyond the scope of this text and the reader is referred to *The Geology of China* by Yang et al. (1986), *Geotectonic Evolution of China* by Ren et al. (1987), *Tectonics of China* by Chen (1989), the *Atlas of the Paleogeography of China* (H.Z. Wang 1985), and the *Tectonic Map of Asia with Explanatory Notes* (Li et al. 1982) for recent and comprehensive reviews of Chinese geology.

2.1 Major Orogenic Events

Table 2.1 summarizes the main orogenic events recognized in China and their approximate correlations with those of North America and Europe. This table is taken from Yang et al. (1986) and, following these authors, it distinguishes between "tectonic stages" and "orogenic movements". The "movements" are phases of intense folding, faulting, and magmatism that may be equated with the terms "orogeny" or "orogenic event". The tectonic "stages" cover much longer periods of time than the "movements" and include phases of quiescence and sedimentation (Yang et al. 1986). Table 2.1 also shows the most important geologic events which are attributed to the various orogenies. The reader should be aware that these "main geologic events" are interpretations which are by no means universally accepted. For example, Yang et al. (1986) and Ren et al. (1987) attribute the Variscan fold belts in northeastern China to a Late Permian convergence of the Sino-Korean Platform with the Siberian Platform. On the other hand, McElhinny

Table 2.1. Chronological table of the major orogenic events in China. (Yang et al. 1986)

Geologic time scale			Tectonic stages		Orogenic movements	Main geologic events	Orogenic movements of Laurasia	
							Europe	N. America
Cainozoic	Quaternary	2.0	Megastage of Pangaea (Gondwana) disintegration	Himalayan Stage	~Himalayan 2~	Upheaval of Qinghai Tibet Plateau	Young Alpedic ~Rodanian~ ~Savian~	
	Neogene	24.6			~Himalayan 1~	Collision of Himalaya and Gangdise		
	Eogene	65			~Yanshanian 3~	Opening of South China sea	~Pyrinean~	
Mesozoic	Cretaceous	144		Yanshanian Stage	~Yanshanian 2~			~Laramidian~
	Jurassic				~Yanshanian 1~	Collision of Gangdise and Qiangtang; Activation of East China continental margin	~Late Cimmerian~ Old Alpedic ~Early Cimmerian~	~Nevadian~
		213						
	Triassic	248	Megastage of Pangaea (Laurasia) formation	Indonisian Stage	~Indonesian 2~ ~Indonesian 1~	Convergence of Yangzi and N. China to form Palasia	~Pfalzian~	
Late Palaeozoic	Permian	286		Hercynian (Variscan) Stage	~Yiningian[a]~	Convergence of N. China and Siberian–Mongolia; Disruption of W. border of Yangzi Platform	Hercynian ~Sudetian~	~Alleghenian~
	Carboniferous	360			~Tianshanian~			
	Devonian	408			~Qilianian~	Formation of S. China Caledonides and close of Quilian troughs	~Bretonian~ ~Erian~	~Acadian~
Early Palaeozoic	Silurian	438						~Taconian~
	Ordovician	505		Caledonian Stage	~Gulangian~		Caledonian ~Salairian~	
	Cambrian	600			~Xingkaian~	Formation of Junggar and other median massifs		
Late Proterozoic	Sinian	850			~Chengjiangian~ ~Jinningian~		~Assyntian~	
	Qingbaikouan	1050	Megastage of platform formation	Jinningian Stage		Formation of Yangzi Platform and Qaidam Massif etc.	~Gothian~	~Grenvillian~
Middle Proterozoic	Jixianian	1400			Sibaoan			
	Changchengian	1850		Luliangian Stage	~Zhongyuean~ ~Luliangian~ (Zhongtianoan)	Formation of N. China and Tarim Protoplatforms	~Karelian~	~Hudsonian~
Early Proterozoic	Hutuoan	2200–2300			~Wutaian~			
	Wutaian	2500–2600			~Fupingian~		~Belomorian~	~Kenoran~
Archean	Fupingian	2900–3000	Megastage of continental nuclei formation	Fupingian Stage		Formation of Ordos and Jilu Nuclei		
		3800						
	Hadean	4500						

[a] Or Nilkaan.

et al. (1981) presented paleomagnetic evidence which indicated that the Sino-Korean and the Yangtze Platforms were widely separated from the Siberian platform and from each other in the Late Permian.

The following discussion of the major orogenic events in China follows the tectonic interpretations of Yang et al. (1986) and Ren et al. (1987) except where other references are given. We use the term "orogeny" and the subordinate term "event" in the same sense as "movement" in Table 2.1 as explained above. Emphasis is given to the Sino-Korean Platform and adjacent areas of northeastern China. For details of the tectonic develop-

18

ment of other areas of China the reader is referred to Ren et al. (1987) and the references given therein.

2.1.1 Archean: Qianxi Orogeny

The earliest orogeny recognized in China is the Early Archean (pre-3 Ga) Qianxi Orogeny (Ma and Wu 1981). The evidence for this is mainly given by Early Archean isotopic ages; the geologic evidence has not survived the deformation and granulite-facies metamorphism of the 2.5 Ga Fuping Orogeny (see below), which affected all known Archean areas in northeastern China.

Little can be said for sure of the nature of the Qianxi Orogeny because of the intense 2.5 Ga overprint. Indeed, Ren et al. (1987) do not recognize a tectonic event at all prior to the Fuping Orogeny, although they acknowledge the presence of rocks older than 3 Ga. Ma and Wu (1981) attribute the granulite-facies regional metamorphism in eastern Hebei province to the Qianxi Orogeny, but Sm-Nd dating by Jahn et al. (1987) and Jahn and Zhang (1984) suggest that the granulite-facies metamorphism, at least in the area north of the Luanhe River (Fig. 1.3), is more likely of Fupingian age (2.5 Ga).

The isotopic ages of 3.4–3.5 Ga reported by Jahn et al. (1987) and Huang et al. (1986) from amphibolite enclaves in orthogneisses south of the Luanhe River are interpreted as the age of mantle-derived basaltic magmatism. Liu et al. (1990) concluded that a sialic basement existed prior to 3.6 Ga based on a $^{207}Pb/^{206}Pb$ zircon age from quartzite. Although it is certain that pre-Fupingian metamorphism and tectonism affected these Early Archean supracrustal rocks, the geologic evidence for such events has apparently been largely destroyed.

2.1.2 Early to Middle Proterozoic: Fuping and Wutai Orogenies

The Fuping Orogeny at 2.4–2.6 Ga involved widespread granulite-facies metamorphism with intense ductile deformation, the intrusion of granitic plutons, and mafic volcanism. Most of the Fuping magmatic rocks which have been studied in eastern Hebei province have near-primitive Sr and Nd isotopic signatures and are interpreted to represent new additions to the continental crust (Pidgeon 1980; Jahn and Zhang 1984; K.Y. Wang et al. 1985). The Fuping Orogeny affected all the known Early Archean rocks of northeastern China, and most isotopic ages from the Archean exposures fall in the "Fupingian" range (Sun and Lu 1985; Yang et al. 1986; Jahn and Ernst 1990). The typical Fuping structural style, according to Ma and Wu (1981), involved the formation of gneiss domes (such as in the Qianan region of eastern Hebei province, see Chap. 2.2.1.1) surrounded by tightly folded belts. The structural trends of the Archean rocks in eastern Hebei province imparted by the Fuping Orogeny are NNE to NE (Ma and Wu 1981).

The Wutai Orogeny (2–2.2 Ga) was a time of major accretion of marginal fold belts and separate continental blocks to form the Sino-Korean Platform (Ren et al. 1987). The Wutai Orogeny is best expressed in fold belts along the margins of the Archean "continental nuclei", where Middle Proterozoic rocks are metamorphosed to lower amphibolite-facies assemblages. In eastern Hebei province, the Wutai Orogeny is represented by a "tectono-thermal event" at about 2.2 Ga resulting in amphibolite-facies metamorphism and NE–SW-trending open folds (Sun 1984). In contrast to the ductile deformation of the earlier orogenies, deep-seated fault zones associated with plutonic activity formed during the Wutai Orogeny (Ma and Wu 1981), testifying to crustal rigidity.

2.1.3 Middle to Late Proterozoic: Zhongtiao and Yangtze Orogenies

The Middle Proterozoic (1.7–1.9 Ga) Zhongtiao Orogeny (also named Luliang) caused further consolidation of, and marginal accretion to the Early Precambrian platforms. In northern China a continuous "proto-continent" was formed by the joining of the Sino-Korean and Tarim Platform (see Fig. 1.1). Local extension of the platform in northern China was marked by the intrusion of extensive mafic dike swarms, anorthosite, and rapakivi granite suites and by the formation of sedimentary rift basins (Ma and Wu 1981). One of these structures is the so-called Yanshan platformal fold belt or Yanshan Trough (also referred to in the literature as the Yanshan settling belt, Yanshan geosyncline, and Yanshan aulocogen) in the northern part of eastern Hebei province. This structure is of relevance to this study because it is host to most of the gold districts described in Chapter 3. The location is marked by the letter B in Fig. 1.1. It is an E–W-trending zone marked by thick accumulations of Middle and Late Proterozoic quartzites and flysch-like sedimentary rocks with K-rich mafic and intermediate volcanic rocks in the basal parts of the sequence (Ma and Wu 1981; Sun and Lu 1985).

The Late Proterozoic Yangtze Orogeny mainly affected southern and southeastern China. The Yangtze Orogeny was important in the consolidation of the Yangtze Platform, and it is commonly subdivided into the Jinning (850 Ma) and Chengjiang (700 Ma) Events (Ren et al. 1987). In northern China evidence for the Jinning Event is found along parts of the southern margin of the Tarim and Sino-Korean Platforms; the Chengjiang Event is not expressed.

2.1.4 Paleozoic: Caledonian and Variscan Orogenies

The main Caledonian orogenic events in China took place in the Late Ordovician to Early Devonian periods, although an Early Caledonian "Xingkai Event" of Middle Cambrian age is recognized in west-central China and in the Indochina peninsula (Ren et al. 1987). The most important areas of Caledonian activity are in southern and western China, where extensive accretionary fold belts represent the addition of eugeosynclinal material (island arcs and marine marginal basins) to the Yangtze and Tarim platforms. Along the northern border of China and in Mongolia, Late Caledonian (Silurian) subsidence and folding are recorded, but here the main phase of orogenic activity was Variscan, when the final collision of the Sino-Korean platform with the Siberian platform took place (according to Ren et al. 1987).

The Variscan Orogeny (Devonian to Permian) involved the collision of the "North China continent", i.e., the Tarim and Sino-Korean Platforms, with the Siberian-Mongolian platform. Variscan fold belts are developed extensively in northwestern and northeastern China, but they are also represented on the southern margin of the Yangtze platform in southeastern China. The well-known Tianshan and Altay mountain ranges in western China are considered typical of the Variscan fold belts by Ren et al. (1987). Several phases of Variscan Orogeny are distinguished by Ren et al. (1987), to which the reader is referred for details. In northeastern China the Variscan Orogeny produced, in addition to fold and thrust belts, large amounts of dioritic to granitic intrusions and minor ultramafic intrusions, which tend to form along major fault zones and lineaments. These rocks are concentrated around the northern and southern borders of the Inner Mongolian Axis in the Sino-Korean Platform (marked by A in Fig. 1.1), and farther to the northeast. Variscan volcanic rocks are not abundant, and only minor intermediate lavas and pyroclastics are known in Permian strata. The volcanic rocks include andesites, pyroxene-bearing andesites, crystal and vitric tuffs and andesitic breccias which formed in the so-called Inner Mongolian eugeosyncline (Yang et al. 1986).

2.1.5 Mesozoic: Indosinian and Yanshanian Orogenies

The Indosinian Orogeny (Triassic) had only minor effects within the Sino-Korean Platform, but was important in southwestern and central China (Fig. 1.1). A narrow zone of Triassic folding and magmatism in central China represents convergence of the Sino-Korean and the Yangtze Platforms (Yang et al. 1986). According to these authors, eastern China was consolidated into a single continental mass by the Indosinian Orogeny. Westward subduction of oceanic lithosphere of the Izanagi plate beneath the eastern margin of the Sino-Korean and Yangtze platforms began in the Late

Triassic. In southwestern China and in the adjacent parts of Indochina, the Indosinian Orogeny is represented by extensive fold belts resulting from the northward convergence of Tethyan island arcs with the Asian continent (Ren et al. 1987).

The Yanshanian Orogeny (Jurassic to Cretaceous) produced wide tectono-magmatic belts on the margins of the Chinese platforms in both eastern China and southwestern China due to continued subduction of oceanic lithosphere (Pacific and Tethyan, respectively) which had begun in the Indosinian Orogeny.

In eastern China, the Yanshanian Orogeny produced a fundamental change in the orientation of structures within the continent from dominantly E–W (related to the N–S accretion and collisions of continental blocks) to dominantly NNE–SSW (caused by interaction with the Pacific margin). The orogeny involved extensive volcanism, plutonism, and the development of elongate NNE-trending fault-bounded basins to a distance of over 1000 km inland from the continental margin. Along the southeastern coastal region of China and the island of Taiwan, regional metamorphism of greenschist grade developed, but elsewhere in eastern China the only metamorphic effects of the Yanshanian Orogeny were contact metamorphism related to igneous intrusions (Yang et al. 1986). The Yanshanian was also the period of most intense metallogenetic activity in eastern China, as discussed more fully in Chapter 2.4.

In western and southeastern China the Yanshanian Orogeny developed during the Late Cretaceous with the formation of extensive fold belts and widespread volcanic and plutonic activity, and with regional uplift. The events reflect continued subduction of the Tethyan oceanic crust and island-arc accretion to the Asian continent (Yang et al. 1986; Ren et al. 1987).

The subdivision of the Yanshan Orogeny is still debated. According to most authors, the Yanshan Orogeny spans the time from Early Jurassic to Late Cretaceous, and can be divided into three phases, as shown on Table 2.1 (Yang et al. 1986; Ren et al. 1987). According to this classification, the third and last phase of the Yanshanian Orogeny terminated in the Early Tertiary. However, Wan and Zhu (1989) argued that the regional stress orientation of the orogenic phase beginning in the Late Cretaceous was different from that of the earlier Yanshanian phases. They suggested that the term Yanshanian Orogeny be limited to the Jurassic, and that the Cretaceous events should be called the Sichuan Orogeny. This suggestion has not yet found wide acceptance and is not followed in this text.

2.1.6 Cenozoic: Pacific Margin and the Himalayan Orogeny

The Cenozoic orogenic events involved, on the one hand, continuing subduction-related magmatism and associated crustal extension at the Pacific margin in eastern China, and on the other, the collision of the Indian subcontinent with Eurasia (Himalayan Orogeny), which affected mostly southwestern and central China. The Himalayan Orogeny produced dominantly compressional features and tremendous regional uplift of the Qinghai-Tibet plateau. Within the Asian continent to the north of the Himalayas, reactivation of earlier mountain ranges and major strike-slip faults developed due to the post-collisional indentation of India into Asia (Molnar and Tapponier 1975).

In eastern China, deep Tertiary and Quaternary sedimentary basins developed in response to crustal extension. Tectonic features are dominated by NNE-trending extensional and strike-slip faults. The opening of the Sea of Japan and of the South China Sea due to back-arc spreading took place from the end of the Cretaceous to the Late Tertiary (Uyeda and Miyashiro 1974; Dickinson 1979). The Bohai Sea and the Yellow Sea are Quaternary features caused by subsidence related to crustal thinning and extensional faults. Cenozoic magmatic activity in eastern China is small in scale and is dominated by mafic volcanism.

2.2 The Precambrian Basement

The Sino-Korean platform contains nearly all of the known exposures of Archean and Early Proterozoic rocks in China (Yang et al. 1986). These exposures consist of high-grade metamorphic rocks of igneous and sedimentary origin, and they represent the ancient crystalline basement of the platform. Middle and Late Proterozoic rocks in China are widespread not only in the Sino-Korean Platform but also in the Tarim Platform to the west, and in the Yangtze Platform to the south (Yang et al. 1986). In the other platforms, the Middle Proterozoic constitutes the crystalline basement but in northeastern China the Middle- and Late Proterozoic rocks comprise weakly metamorphosed platform sequences of carbonate rocks and clastic sedimentary rocks. The distribution of Precambrian exposures in northeastern China is shown in Fig. 2.1 based on the *Tectonic Map of the People's Republic of China* at a scale of 1:4 000 000 (Chinese Academy of Geological Sciences 1979). The reader is referred to Yang et al. (1986), Sun and Lu (1985), and Ma and Wu (1981) for general reviews of the Precambrian of China.

It was shown in Chapter 1.3 that, in eastern Hebei province, the Archean rocks host most of the gold deposits (see also Fig. 1.3). Therefore the emphasis of the chapter is on a description of the Archean rocks. The Early Proterozoic rocks in eastern Hebei province host some minor gold deposits and occurrences, for example the Banbishan deposit described in Chapter 3.5. The Late Proterozoic platform rocks are rarely hosts to gold deposits, although a notable exception is the Yuerya gold deposit, which is described in Chapter 3.3.

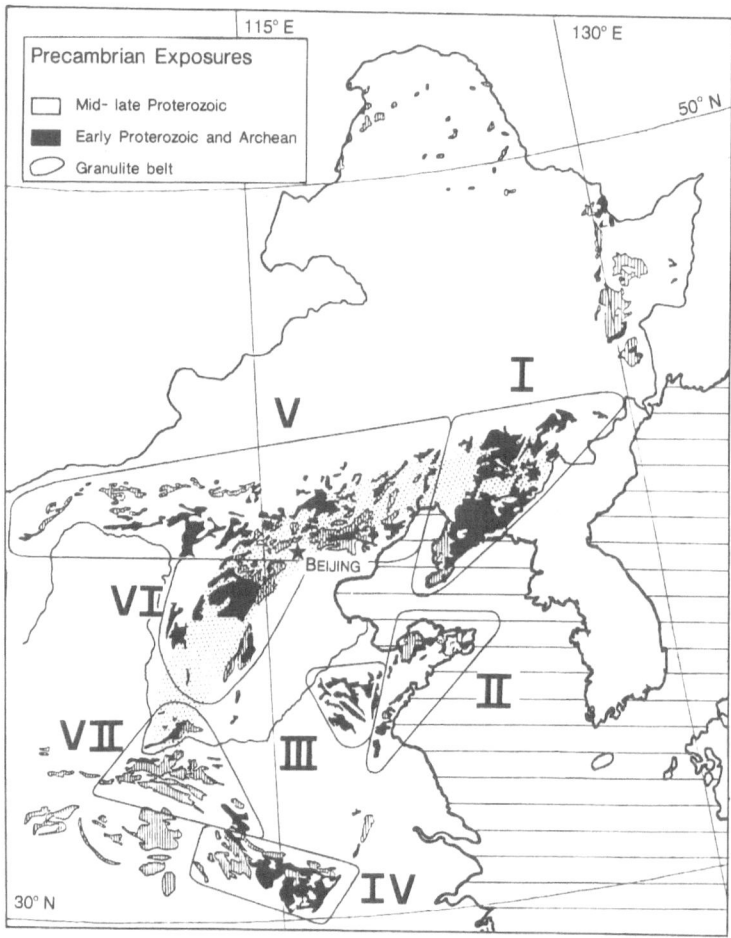

Fig. 2.1. The distribution of Precambrian outcrops in northeastern China, from the Chinese Academy of Geological Sciences (1979) 1:4000000 scale tectonic map of China. *Roman numerals* refer to the Precambrian massifs listed in Table 2.2

2.2.1 Archean Rocks

The Precambrian rocks exposed in eastern Hebei province represent nearly 3000 million years of earth history, and the province contains the best-studied Archean rock sequences of the Sino-Korean Platform. The oldest known rocks are Early Archean granulite-facies metasedimentary rocks in the Qianan area from which zircons have recently been dated at 3600 Ma (Liu et al. 1990). A wealth of data on the lithology, mineral composition, whole-rock chemistry, and radiogenic isotopic composition of the Archean rocks has been published, and research on these rocks is very active (Sun and Wu 1981; Zhang and Cong 1982; Jahn and Zhang 1984; Sun 1984; K.Y. Wang et al. 1985, 1990; R.M. Wang et al. 1985; Huang et al. 1986; Jahn et al. 1987; Sills et al. 1987a; Jahn 1990, 1991; Liu et al. 1990). Before the rocks of eastern Hebei province are discussed in detail, it is worth reviewing some general features of the Archean geology in northeastern China together with observations from other areas.

Regional Comparisons

Yang et al. (1986) distinguished seven areas in northeastern China where Archean rocks are exposed, and these are marked in black in Fig. 2.1. In all of these, the Archean is characterized by complex, polymetamorphic sequences of high grade supracrustal rocks, frequently showing evidence of migmatization and intrusion by one or more generations of felsic plutonic rocks now represented by orthogneisses. It is impossible to understand the evolution of these rock complexes without abundant isotopic age data, and these data are only beginning to become available in some areas. For this reason, the correlations among the separate Archean complexes are tentative, and even within a single complex several schemes of nomenclature and subdivision may exist. Table 2.2 lists the correlation and nomenclature of the Archean and Early-Proterozoic lithostratigraphic groups in northeastern China according to Yang et al. (1986).

According to Yang et al. (1986), all the Archean rocks in northeastern China have been metamorphosed to at least the upper amphibolite facies, and most show evidence of anatexis. Retrograde greenschist facies and lower amphibolite facies assemblages are common. Granulite-facies assemblages are found in three of the seven areas, and the distribution of

Table 2.2. Correlation of rock units in seven Early Precambrian exposures in northeastern China. (After Yang et al. 1986)

Region	I	II	III	IV	V	VI	VII
	Liaoning and Jilin	Eastern Shandong	Western Shandong	Huaiyang	Yanshan	Wutai Taihang	Eastern Qingling
Early Proterozoic	Liaohe Group	Fenzishan Group		Hongan Group	Zhuzhangzi Group	Wutai Group	Songshan Group
Archean	Anshan Group	Jiodong Groupd	Taishan Group	Dabie Group	Dantazi Group	Longquanguan Group	Dengfeng Group
					Qianxi Group	Fuping Group	

these led Yang et al. (1986) to propose the concept of a "northern granulite belt" (stippled in Fig. 2.1).

In spite of the uncertainty caused by intense metamorphism, the Archean rocks can be divided into two categories based on their protoliths: supracrustal rocks (metavolcanic and metasedimentary) and plutonic rocks (orthogneisses). Most of the Archean supracrustal rock types present include amphibolites, pelitic (garnet-sillimanite) gneisses, magnesian marbles and calc-silicate gneisses, metamorphic banded iron formations, quartzites, graphitic schists, and pelitic schists. The common Archean gray felsic gneisses, mainly biotite-plagioclase gneisses or hornblende-plagioclase gneisses, are of ambiguous origin. In some cases, field relations suggest that the felsic gneisses, in particular the granoblastic ones, were intercalated with the supracrustal rocks and may represent immature arc sediments or tuffs (Liu et al. 1990). In other cases there is no good evidence of the protolith character.

The plutonic rocks, now represented by orthogneisses, have a wide range of granitic compositions from tonalite to granite. Where the rocks have been studied in detail, for example in Liaoning province (Ernst et al. 1988; Wang et al. 1990) and in eastern Hebei province (Liu et al. 1990), it was found that there were several generations of intrusion. Early intrusions contain enclaves of supracrustal rocks (amphibolites, aluminous gneisses). These early orthogneisses resemble the TTG (trondhjemite-tonalite-granodiorite) gneisses common in Archean terranes throughout the world (Windley 1984). Both the TTG orthogneiss and their enclaves have been metamorphosed at the upper amphibolite to granulite facies. Later intrusions, which formed after the peak of high-grade metamorphism, are of more granitic composition (granodiorite-quartz monzonite-granite).

The relative proportions of the supracrustal rocks and orthogneisses were roughly estimated in eastern Hebei province and in the Taihang-Wutai Mountains area. According to Sun and Wu (1981), the proportion of orthogneisses (their "migmatitic granite") in these two areas is 39 and 42%, respectively, and the supracrustal rock types (amphibolites, metavolcanics, "granulitites", iron formations, and marbles) total about 25% in eastern Hebei and about 30% in the Taihang-Wutai area. In a later study, Wang et al. (1990) state that about one-third of the rocks of the high-grade Archean terrane at Qianan in eastern Hebei province have supracrustal protoliths.

A brief comparison of the Archean geology in northeastern China with that of the Archean greenstone-belt terranes on other cratons is necessary in order to judge the applicability of metallogenetic models for greenstone-belt gold deposits to the Chinese examples (see also Chap. 6). The Archean terrane in eastern Hebei province resembles the "granulite-gneiss belts" of Windley (1984), as exposed in Greenland, Labrador, Scotland, and Limpopo, much more than it does the greenstone belts. Wang et al. (1990) also emphasized the similarity of the Qianxi Group supracrustals (eastern

Hebei province) with the Isua supracrustals of Greenland. The characteristic rock types include marbles, quartzites, possible bimodal gneiss-amphibolite suites, abundant banded iron formations, and pelitic schists. Windley (1984) emphasized that the contrast between the high-grade granulite-gneiss terranes and the greenstone belts is not simply a question of metamorphic grade because there are major differences in their protoliths. One important distinction between a high-grade greenstone belt and a granulite-gneiss terrane, according to Windley, is that the supracrustal rocks in the greenstone belts are typical of the eugeosynclinal environment, with calc-alkaline volcanics, immature graywacke-turbidite sedimentary rocks and Algoma-type iron formations. The granulite-gneiss terranes, on the other hand, contain tholeiitic basalts and shelf sediments including carbonates, quartzites, and pelitic rocks. The latter series of rock types seems to be dominant in most of the Archean sequences in northeastern China according to the descriptions in Yang et al. (1986). However, in contrast to many of the world's granulite-gneiss terranes, the Chinese Archean contains no anorthosite complexes.

There are, in fact, some typical greenstone belt rocks in the Archean of northeastern China but they are very much underrepresented in comparison with the high-grade terranes. Zhai et al. (1985) reported, for example, greenstone/granite-type rocks in the Archean Qingyuan terrane of northern Liaoning province with basal units of mafic and ultramafic rocks overlain by calc-alkaline metavolcanics and eugeosynclinal metasediments, and surrounded by diapric granites.

2.2.1.1 Lithologies in Eastern Hebei Province

It is unfortunate that two different systems of subdivision and nomenclature are currently in use for the Archean and Early Proterozoic rocks in eastern Hebei province. These are compared in Table 2.3. The nomenclature of Sun Dazhong and coworkers of the Tianjin Institute of Geology, Chinese Academy of Geological Sciences (Sun 1984; Sun and Lu 1985) is best known in international publications. However, the authors of this text decided, after some debate, to use the alternative classification of the Ministry of Metallurgical Industry outlined in Table 2.3 because it is more consistent with the usage in the gold districts of eastern Hebei province.

Both systems of nomenclature agree on the subdivision of the Middle- and Late Proterozoic. There are major differences in the terminology of the Early Proterozoic and Archean units and, unfortunately, it is not possible to offer an exact correlation between the two systems, partly because of large differences in the reported thickness of the various units and partly because of the paucity of radiometric ages. For example, the Archean section (pre-2500 Ma) is about 30 km thick according to Q.S. Zhang et al. (1984), whereas Sun (1984) considers the total Archean section to be on the order of only

Table 2.3. Correlation of Early Precambrian rock units in eastern Hebei province

Sun (1984) Fm.	Sun (1984) System/Group	This study System/Group	This study Fm.
Jingeryu Fm.	Qingbaikou System	Qingbaikou System	Jingeryu Fm.
Xiamaling Fm.			Xiamaling Fm.
Tieling Fm.	Jixian System	Jixian System	Tieling Fm.
Hongshweizhuang Fm.			Hongshweizhuang Fm.
Wumishang Fm.			Wumishang Fm.
Yangzhuang Fm.			Yangzhuang Fm.
Dahongyu Fm.	Changcheng System	Changcheng System	Gaoyuzhuang Fm.
			Dahongyo Fm.
Changzhougou Fm.			Tuangshanzi Fm.
Boluotai Fm.	Qinglonghe Group		Changlingou Fm.
			Changzhougou Fm.
Zhangjiagou Fm.		Zhuzhangzi Group	Zhalangzhangzi Fm.
Xiabaicheng Fm.	Shuanshanzi Group		Shangbaichengzi Fm.
Luzhangzi Fm.			Zhangjiagou Fm.
Ciyushan Fm.			Luzhangzi Fm.
Sanmendian Fm.	Badaohe Group	Dantazi Group	Sanhedian Fm.
			Nandianzi Fm.
Wanzhangzi Fm.			Fenghuanzui Fm.
Wangchang Fm.			Baimaozi Fm.
		Qianxi Group	Malanyu Fm.
Santunying Fm.	Qianxi Group		Santunying Fm.
Shangchuang Fm.			Shangchuang Fm.

10 km thick. The Early Proterozoic section, according to Q.S. Zhang et al. (1984), totals about 9.5 km in thickness, whereas the time-equivalent section of Sun and coworkers is about 4.5 km thick. This mismatch in the reported thickness of the Early Precambrian units undoubtedly reflects the intense deformation of these rocks. Indeed, Jahn and coworkers have criticized the

usefulness of treating the Qianxi Group in a stratigraphic sense at all (Jahn and Zhang 1984; Jahn et al. 1987; Liu et al. 1990). Liu et al. (1990) recommend that the term Qianxi Group be abandoned in favor of "Qianxi Complex" because of the heterogeneity of the rocks in terms of both lithology and age. We retain the older "stratigraphic" usage here, but note that the definition of the Archean units in northeastern China is in a state of flux.

The following descriptions of the rocks are based in part on published reports and in part on our own observations in outcrops north of the Luanhe River near the towns of Santunying, Qianxi and Qinglong, and in the mining districts of Niuxinshan, Sanjia, and Jinchangyu.

Early Archean: Supracrustal Rocks and Orthogneisses

The Early Archean rocks in eastern Hebei province include metamorphosed supracrustal rocks assigned to the Qianxi Group and orthogneisses which are intrusive into, and include enclaves of, the Qianxi Group rocks. The orthogneisses are not formally included within the Qianxi Group according to current usage, and are therefore described separately below. The Qianxi Group is divided into three formations in ascending order as follows: Shangchuang, Santunying, and Malanyu Formations (see Table 2.3).

Shangchuang Formation

Rocks of granulite facies predominate in the lowest formation of the Qianxi Group, and both mafic and felsic varieties are common. The mafic granulites are granoblastic, two-pyroxene-quartz-plagioclase rocks with or without minor garnet and hornblende. Felsic granulites generally lack orthopyroxene and comprise clinopyroxene, plagioclase, quartz, and minor garnet. Most granulites show extensive retrograde metamorphism to amphibolite facies assemblages, and in addition, many rocks are migmatized with veins and/or schlieren of granitic mobilizates. Minor rock types in the Shangchuang Formation include ultramafic rocks (hornblendites, hornblende-pyroxenites) which may represent mafic dikes, and banded hornblende-magnetite quartzites or magnetite-rich amphibolites interpreted to be metamorphic iron formations. The magnetite-rich rocks are locally mined for iron ore. Detailed study of the mafic and felsic granulites by Jahn and Zhang (1984) suggests that most of these have an igneous parentage with a tholeiitic and calc-alkaline affinity (Chap. 2.2.1.2).

Santunying Formation

This unit contains amphibolites, granulites, and metasedimentary gneisses. There are fewer granulite-facies rocks in the Santunying Formation than in the Shangchuang Formation. The most common rock types are hornblende-plagioclase amphibolites with or without biotite, banded magnetite quartzites, quartz-magnetite amphibolites, and granoblastic felsic biotite gneisses (so-called leptites). The Santunying Formation is an important source of iron ore in eastern Hebei province. Metapelitic garnet-sillimanite-biotite-plagioclase gneisses, calc-silicate gneisses, and quartzites occur locally. Graphite schists occur at the locality of Qianzhuang in Qinglong county. Retrograde metamorphic textures and mineral assemblages are common in all of the rocks and migmatisation is extensive.

Malanyu Formation
Granulite-facies rocks are rare in the Malanyu Formation and migmatization is not widespread. The main rock types are layered amphibolites, garnet amphibolites, banded pyroxene-hornblende-plagioclase gneisses and schists, and feldspathic biotite schists. Lenses of magnetite-rich amphibolite and magnetite quartzite are also present.

Orthogneisses
Orthogneisses of Archean age in eastern Hebei province are concentrated in the area around Qianan, south of the Luanhe River, and in the easternmost part of the province near the cities of Qinghuangdao and Shanhaiguan (see Fig. 1.3). Orthogneisses are far less common in the area where gold deposits are concentrated, i.e., north of the Luanhe River and west of the Shanhaiguan area. This irregular distribution of orthogneisses and supracrustal rocks has led some workers to divide the Archean terrane of eastern Hebei into two geologic domains (R.M. Wang et al. 1985; Jahn et al. 1987). The supracrustal domain north of the Luanhe River is referred to as the Qianxi highly folded domain or Zunhua Belt, and the area south of the Luanhe River is termed the Qianan Complex or the Qianan Gneiss Dome. These workers did not extend their classification to the Shanhaiguan area, which would logically constitute a third geologic domain.

The orthogneisses of the Qianan area have been described by K.Y. Wang et al. (1985) and by Liu et al. (1990). According to these authors, the gneisses comprise two main compositional and age groups. The first group includes tonalitic to granodioritic gneisses, showing clear intrusive relationships with the supracrustal rocks of the Qianxi Group. These rocks have been metamorphosed together with their supracrustal enclaves at upper amphibolite or granulite facies conditions. The second group of Archean orthogneisses comprises weakly foliated charnockites and granites which intrude both the early orthogneisses and the Qianxi Group metamorphic rocks. The composition of these rocks, summarized in Chapter 2.2.1.2, ranges from tonalitic to granitic, but is significantly more siliceous than the early orthogneisses. K-metasomatism of the country rocks is often observed around the Late Archean intrusions.

The orthogneisses of the Shanhaiguan area have been divided into three intrusive complexes, namely the Anziling Migmatitic Granite, the Jialingkou Quartz Diorite, and the Shanhaiguan Granite. These complexes, especially the Anziling Migmatitic Granite, contain supracrustal relics including iron formations. Field observations and isotopic age dating indicate that the orthogneisses intruded after the regional metamorphism of the Dantazi Group, i.e., in the Late Archean or Early Proterozoic (Wang et al. 1987).

Late Archean: the Dantazi Group

The Late Archean rocks in eastern Hebei province are assigned to the Dantazi Group. The rock types of the different formations in the Dantazi Group are summarized below in ascending order. Correlations with the nomenclature of Sun (1984) are shown in Table 2.3.

Baimiaozi Formation
This formation contains mainly granoblastic biotite-plagioclase gneisses, hornblende-biotite-plagioclase gneisses and, in some districts, layers of actinolite magnetite quartzite and cummingtonite-grunerite-magnetite quartzite. The latter locally forms an important source of iron ore.

Fenghuanzui Formation
The Fenghuangzui Formation contains mainly amphibolites with minor hornblende-plagioclase gneisses, leptites with marble lenses, and magnetite-rich hornblende quartzites.

Nandianzi Formation
The Nandianzi Formation consists of biotite-plagioclase gneisses, biotite and chlorite schists, amphibolites, and marble lenses.

2.2.1.2 Chemical Composition and Protolith Rock Types

Qianxi Group
The chemical compositions of meta-igneous rock types of the Qianxi Group (amphibolites, mafic and intermediate-felsic granulites) have been reported by Jahn and Zhang (1984), R.M. Wang et al. (1985), and Wang et al. (1990). According to the studies cited above, the mafic rocks have major element and REE compositions similar to both continental tholeiite basalts and island arc tholeiites. The intermediate and felsic rocks have a calc-alkaline affinity. However, attempts to characterize these rocks in terms of volcanic series and tectonic setting are hampered by the uncertainty introduced by the effects of later geologic events, and any conclusions should be considered with caution. Some aspects of the composition of these rocks taken from the literature are shown in discrimination diagrams in Fig. 2.2.

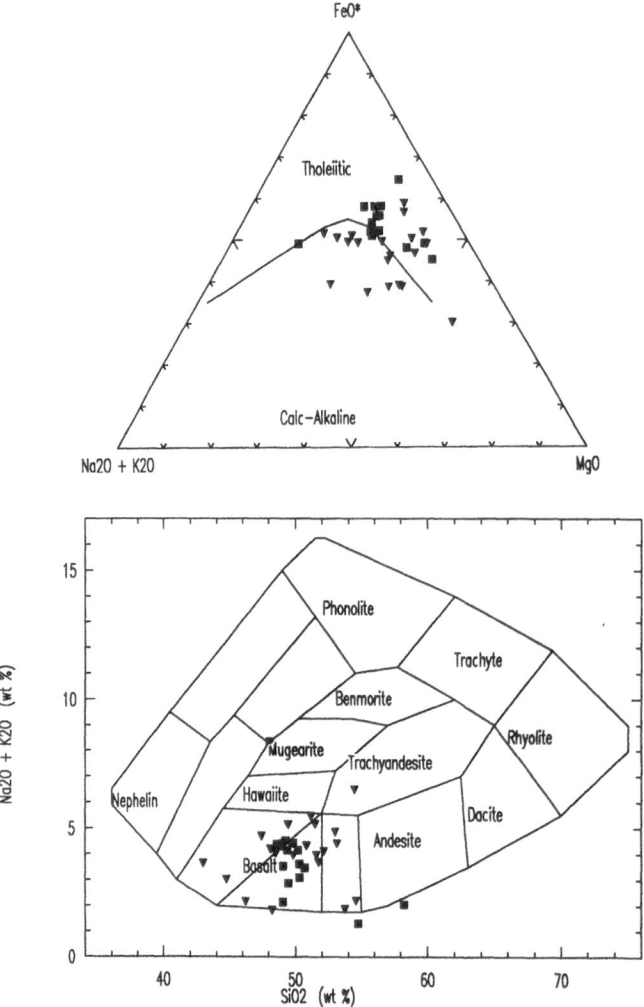

Fig. 2.2. AFM and alkali-silica diagrams showing the compositional range of amphibolites (*triangles*) and mafic granulites (*squares*) from the Qianxi Group. The data plotted were taken from Jahn and Zhang (1984) and Wang et al. (1990)

Orthogneisses

The chemical composition of orthogneisses of the Qianan area have been documented by K.Y. Wang et al. (1985) and by Liu et al. (1990). The early generation of tonalite-granodiorite orthogneisses have major and trace element compositions similar to the calc-alkaline trondhjemite-tonalite-granodiorite (TTG) gneisses typical of other Archean high-grade regions (Wang et al. 1990). The normative feldspar compositions of these rocks are shown in Fig. 2.3. The charnockites have compositions similar to the

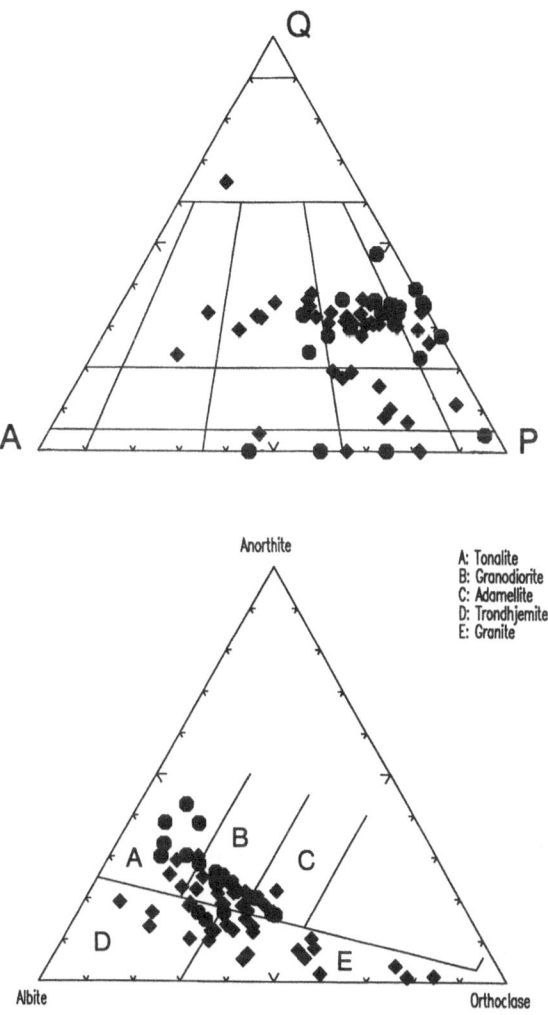

Fig. 2.3. Streckeisen QAP and normative feldspar diagrams showing the compositional range of Archean orthogneisses (*octagons*) and charnockites (*diamonds*) in eastern Hebei province. The data plotted were taken from K.Y. Wang et al. (1985) and Wang et al. (1990)

TTG-orthogneisses, although their bulk compositions tend to be more granitic. Although they formed at different times (see below), both series of orthogneisses and charnockites are thought to have most likely formed by partial melting of the mafic granulite and amphibolite basement (R.M. Wang et al. 1985).

Dantazi Group

There are no compositional data available for the rocks of the Dantazi Group, and none were obtained in our investigations. According to the rock types present, the protoliths are interpreted to be mafic and intermediate volcanic rocks in the lower part, grading upward to dominantly intermediate volcano-sedimentary and sedimentary rocks in the upper part (Yang et al. 1986).

2.2.1.3 Isotopic Age

The oldest rocks of the Qianxi Group presently known are supracrustal enclaves in orthogneisses in the Qianan area. The oldest age is a $^{207}Pb/^{206}Pb$ single zircon model age of 3.65 Ga obtained by Liu et al. (1990) from a fuchsite-bearing quartzite. This clearly suggests the existence of a sialic basement older than 3.6 Ga in northeastern China. Amphibolite enclaves in the Qianan orthogneisses have yielded ages of 3.4–3.6 Ga by the Sm-Nd method (Jahn and Zhang 1984; Huang et al. 1986; Jahn et al. 1987). These Sm-Nd ages are interpreted as the crystallization ages of the protoliths. Interestingly, the amphibolites and granulites from the Qianxi Group north of the Luanhe River have not yielded any reliable ages older than 3.0 Ga, which perhaps strengthens the argument that two different Archean domains are present in eastern Hebei. The granulites in the Shangchuang formation of the Qianxi Group north of the Luanhe River have been dated by Rb-Sr, Sm-Nd, and zircon U-Pb methods by different laboratories and all give ages around 2.5 Ga (Pidgeon 1980; Jahn and Zhang 1984). Jahn and Zhang (1984) concluded from their Sm-Nd and Rb-Sr data that the 2.5 Ga ages reflect the age of granulite-facies metamorphism, but that the protolith basalts cannot have formed much more than 100 Ma prior to metamorphism. The orthogneisses intrusive into the Qianxi Group in the Qianan area have been dated by U-Pb and Pb-Pb methods on zircon, as summarized by Liu et al. (1990). The results show a range of ages between about 2.9 and 2.5 Ga with a pronounced cluster at about 2.5 Ga. The ages were all interpreted as crystallization ages. K.Y. Wang et al. (1985) reported a Rb-Sr whole rock isochron age of 2650 ± 50 Ma from charnockites from the same area, which they also interpreted as the crystallization age of the rocks. The orthogneisses in the Shanhaiguan area seem to be younger based on their Rb-Sr ages of 2412–2446 Ma (Wang et al. 1987).

Figure 2.4 summarizes the published isotopic age data from the Archean rocks in eastern Hebei. The earliest supracrustal rocks formed at about 3.6 Ga and these rocks preserve evidence of a still older sialic basement. Extensive intrusion of tonalite-trondjhemite rocks at about 2.5 Ga in the Qianan area was broadly contemporaneous with granulite facies regional metamorphism. The 2.5 Ga events, referred to as the Fuping Orogeny (Chap. 2.1.2), are reflected in isotopic ages from other areas of northeastern

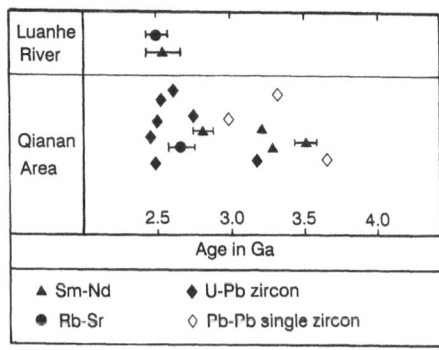

Fig. 2.4. Summary of isotopic age data from Archean rocks in eastern Hebei province. (After Liu et al. 1990)

China as well. To date, no good evidence of Early Archean rocks (pre-3.0 Ga) in notheastern China outside of eastern Hebei province has been reported. A younger group of ages from the Qianxi Group rocks, not shown on Fig. 2.4, represents later thermal events related to the Wutai (2 Ga) and/or Zhongtiao (1.7 Ga) orogenies. Huang et al. (1986) report a Sm-Nd whole rock isochron age of 1.7 Ga for a quartz diorite whose depleted mantle model age is 3.4 Ga. Jahn and Zhang (1984) obtained a Rb-Sr mineral isochron age of 1.68 Ga from granulites whose whole-rock age is 2.5 Ga. K.Y. Wang et al. (1985) attributed the Rb-Sr whole rock age of 2.1 Ga for gray gneisses near Qianan to isotopic resetting by the Wutai Orogeny.

Relatively few radiometric age data are available from the Dantazi Group rocks. Yang et al. (1986) cite K-Ar hornblende and whole-rock ages ranging from 2435 to 2660 Ma and one Rb-Sr whole rock age of 2523 ± 139 Ma for the Dantazi Group as a whole. Sun and Lu (1985) reported age data from the correlative rocks of the Badaohe Group (see Table 2.3), including a Rb-Sr whole rock isochron age of 2550 ± 45 Ma from the Wangcheng Formation and an U-Pb zircon age of 2494 ± 24 Ma from migmatitic granite in the Badaohe Group. These were interpreted by Sun and Lu (1985) as the age of high-grade metamorphism and migmatization related to the Fuping Orogeny.

2.2.1.4 Metamorphism

The P-T conditions of granulite metamorphism in rocks of the Qianxi Group have been estimated at 750 to 850 °C and 8 to 11 kbar based on two-pyroxene and garnet-pyroxene equilibria (Zhang and Cong 1982; Sun 1984; Sills et al. 1987a). The presence of sillimanite, cordierite, and garnet in

35

metapelitic horizons in the Qianxi Group suggests medium-pressure metamorphic conditions. Retrograde metamorphism is a common feature of the Qianxi Group rocks, and many of the granulite facies rocks have been transformed into retrograde amphibolites. Sun (1984) reported P-T estimates of retrograde metamorphism at 550–750 °C and 6–10 kbar. R.M. Wang et al. (1985) reported garnet-biotite temperatures of about 600 °C from a retrograded granulite from the Qianan area.

Quantitative estimates of metamorphic conditions in the rocks of the Dantazi Group have not been published. The mineral assemblages of the rocks suggest that the metamorphism reached the upper amphibolite facies and the onset of anatexis. No reliable estimates of pressure can be made from the available information.

2.2.2 Proterozoic Rocks

The Proterozoic Eon in China is defined by Yang et al. (1986) as extending from 2500 to 850 Ma. Between 850 Ma and the beginning of the Cambrian Period Chinese geologists recognize the "Sinian System". The Proterozoic rocks in eastern Hebei province can be usefully divided into two series based on their structural and metamorphic nature. The first is of Early Proterozoic age (ca. 2500–1800 Ma), and it forms a strongly folded and foliated medium-grade metamorphic basement which shares some tectonic elements with the Archean. The second series formed in the Middle and Late Proterozoic (ca. 1800–850 Ma), and it consists of platformal clastic and carbonate sedimentary sequences which are weakly metamorphosed and only gently folded.

The Proterozoic rocks in northeastern China have been much less intensely studied from a geochemical and isotopic standpoint than have the Archean rocks. It is not possible to provide details of the chemical composition, isotopic composition, and P-T conditions of metamorphism as was the case for the Archean rocks. For the present purpose, this lack of information is not critical, since most of the gold deposits in eastern Hebei occur in the Archean rocks. The discussion below is therefore limited to a listing of the Proterozoic formations present in eastern Hebei province and their lithologic makeup, together with such age data as are available. A full discussion of Proterozoic stratigraphy in China is given in Yang et al. (1986) and references therein.

The following lithologic descriptions and age data are taken mostly from information in Yang et al. (1986), Sun and Lu (1985), and partly from unpublished data from the Ministry of Metallurgical Industry. The stratigraphic nomenclature is given in Table 2.3.

2.2.2.1 Early Proterozoic

The Early Proterozoic rocks are poorly represented in eastern Hebei province in contrast to the Archean rocks. In many parts of eastern Hebei province Middle- to Late Proterozoic rocks rest unconformably or tectonically on the Archean, and the Early Proterozoic is entirely missing (see Fig. 1.3). The best-exposed type section of the Early Proterozoic in northeastern China is found in the Wutai–Taihang mountains area west of Beijing (area VI in Fig. 2.1). In this area, the Early Proterozoic (Wutai Group) is characterized by greenschist or lower amphibolite facies flysch-like turbidite sequences with minor BIF layers, and intermediate to felsic volcanoclastic rocks (Yang et al. 1986). Similar rock types occur in eastern Hebei province but their nomenclature is different (see Table 2.2), and correlation between the two areas is tentative.

The Early Proterozoic rocks in eastern Hebei province are represented by the Zhuzhangzi Group, which is exposed in a limited area east of Qinglong (see Fig. 1.3). Five formations of the Zhuzhangzi Group are recognized as described below in ascending order:

Sanhedian Formation
The formation consists of feldspathic sericite-quartzites, sericite-quartz schists, sericite schists, and magnetite-rich quartzites.

Luzhangzi Formation
The dominant rock types include amphibolites with relict pillow structure, hornblende schists, and chlorite-hornblende schists. Isotopic ages (Rb-Sr) reported from the unit are 2217 ± 43 Ma (Lu and Huang 1987) and 2193 Ma (Sun 1984).

Zhangjiagou Formation
The main rock types are meta-conglomerates with volcanogenic clasts and biotite leptites. The isotopic age (Rb-Sr) is 2223 ± 136 Ma (Lu and Huang 1987). Sun (1984) reported a Rb-Sr age of 2398 Ma.

Shangbaichengzi Formation
This formation contains biotite leptites, biotite-hornblende leptites, hornblende schists, and lenses of iron-rich quartzite showing cross-bedding. The isotopic age (Rb-Sr) is 1824 ± 66 Ma (Lu and Huang 1987).

Zhalangzhangzi Formation
This unit includes sericite schists, biotite schists, biotite leptites, and iron-rich quartzites, all locally containing garnet.

2.2.2.2 Middle to Late Proterozoic

The Middle- and Late Proterozoic rocks (1800–850 Ma) are well represented in eastern Hebei province, and one of best exposed type sections in China, some 9.5 km thick, is at Jixian, located a few tens of kilometers west of Qianan (for location, see Fig. 1.3). The rocks consist of weakly deformed, very low-grade metamorphic sedimentary and volcano-sedimentary sequences. The Middle- and Late Proterozoic rocks are divided into three systems; in ascending order these are the Changcheng System, Jixian System, and Qingbaikou System (Table 2.3). A brief description of the formations in each system is given below.

1. The Changcheng System

Changzhougou Formation
The basal formation of the Changcheng system consists of conglomerates and coarse sandstones in the lower part and sandy siltstones in the upper part.

Changlingou Formation
This unit consists of silty slates and shales in the lower and middle parts; and slates, shales and minor carbonaceous dolomites in the upper part. The isotopic age (K-Ar) is 1875–1817 Ma (Zhong 1975).

Tuangshanzi Formation
This unit consists of argillaceous and siliceous dolomites in the lower part, and silty micrite dolomites and dolomitic sandstones in the upper part. The isotopic age is 1776 Ma (U-Pb whole rock, Zhong 1975).

Dahongyu Formation
This formation contains transgressive sandstones and arkoses, flinty micrites and siliceous laminated dolomites. Potassic volcanic rocks occur in the middle part. The isotopic age is 1678 Ma (K-Ar glauconite, Zhong 1975).

Gaoyuzhuang Formation
The formation contains basal quartz arenites, dolomitic sandstones in the lower part, flinty dolomites in the middle part, and an upper part of manganiferous dolomites and micritic dolomitic limestones.

2. The Jixian System

Yangzhuang Formation
The main rock types are silty argillaceous dolomites with dolomitic limestones, and limestones.

Wumishan Formation
The main rock types are dolomites and flinty banded dolomites in the lower part, and dolomitic limestones in the upper part.

Hongshweizhuang Formation
The unit contains dolomites in the lower part, and slates with thin beds of sandstone in the upper part.

Tieling Formation
This formation contains dolomitic limestones, manganiferous limestones and shales in the lower part, and stromatolitic limestones in the upper part. The isotopic age is 1205 to 1132 Ma (K-Ar glauconite, Zhong 1975).

3. The Qingbaikou System

Xiamaling Formation
The Xiamaling Formation contains conglomerates, ferruginous sandstones, and paleosols at the base, which represent an erosional surface on the underlying limestones. The basal part is succeeded by shales and sandstones in the lower part, and shales with siltstones in the upper part.

Jineryu Formation
This formation forms a transgressive sequence with feldspathic conglomerates at the base followed upward by arkosic sandstones and shales in the lower part and dolomitic micritic limestones in the upper part. Yang et al. (1986) reported a K-Ar glauconite age of 899–855 Ma for the formation.

2.2.2.3 Metamorphism and Nature of the Protoliths

The metamorphic grade of the Zhuzhangzi Group rocks is in the greenschist to epidote-amphibolite facies according to the reported mineral assemblages. The protoliths are dominated by volcanic and volcaniclastic rocks, with mafic volcanic agglomerates and lavas at the base, overlain by intermediate to felsic volcanic rocks with upward-increasing volcaniclastic components. The protoliths of the upper formations are dominated by volcaniclastic rocks intercalated with semipelitic sedimentary rocks.

The Middle and Late Proterozoic rocks are essentially unmetamorphosed epicontinental platform sequences representing cyclic sedimentation of clastic and carbonate material. Volcanogenic material is locally present but is rare in comparison to the Early Proterozoic section.

2.3 Structural Geology

Eastern Hebei province is situated in the eastern part of the Sino-Korean platform near the border between two "second-order" tectonic units, namely the Inner Mongolian Axis and the Yanshan Platformal Fold Belt. These are marked in Fig. 1.1 by the letters A and B, respectively. The area has been subject to repeated orogenies, as summarized in Chapter 2.1, the most intense of which were the Fuping (2500 Ma), Zhongtiao (1800 Ma), and Yanshan (200–150 Ma). Particularly the Yanshan Orogeny produced extensive, deep fracture/fault zones, some of crustal scale.

2.3.1 Precambrian Structures

Figure 2.5 shows a simplified tectonic map of Early Precambrian exposures in eastern Hebei province after Sun et al. (1989). The Early Precambrian rocks are exposed in a series of generally E–W-trending anticlinoria. These relatively open folds represent a late stage of folding attributed by most authors to the Zhongtiao Orogeny (ca. 1800 Ma). Folds of an earlier stage (Fuping and pre-Fuping) are tight to isoclinal with nearly N–S axes and west-dipping axial planes (Sun 1984; Sun et al. 1989). As a result of the late stage folding, the strike direction of lineations and foliations in the Archean and Early Proterozoic rocks changes from dominantly NE–SW in the north to N–S in the south.

According to Sun (1984), the area shown in Fig. 2.5 is divided into an eastern and a western tectonic unit by the major NW-trending Lengkou Fracture Zone (marked on the tectonic map). The western unit is uplifted relative to the eastern, thus the Archean rocks are most abundant west of the Lengkou Fracture Zone. In the western tectonic unit early isoclinal folds are well exposed in the Qianxi Group in the Santunying-Taipingzhai region of Qianxi county, at Malanyu in the western part of Zunhua county, and in the western part of Qianan county. Late stage open folding is best represented by the large Malanyu-Taipingzhai Anticlinorium (MTA on the map) which extends E–W through the center of the area.

In the eastern tectonic unit Early Proterozoic rocks are more commonly exposed and there are large areas of migmatitic granite. The early stage of tight folds with N–S axial planes dipping west is present, but the early folds are more open than in the western tectonic unit. On the other hand, the E–W-trending late stage folds are tighter than in the western tectonic unit; an example shown in the tectonic map is the Qianzhangzi-Longwangmiao Anticlinorium (QLA on the map).

The Archean and Proterozoic structures played a major role in influencing later structures, especially the position of intrusions and main fracture zones, both of which are important for localizing gold mineralization. Both the early and Late Precambrian fold generations developed reverse faults in

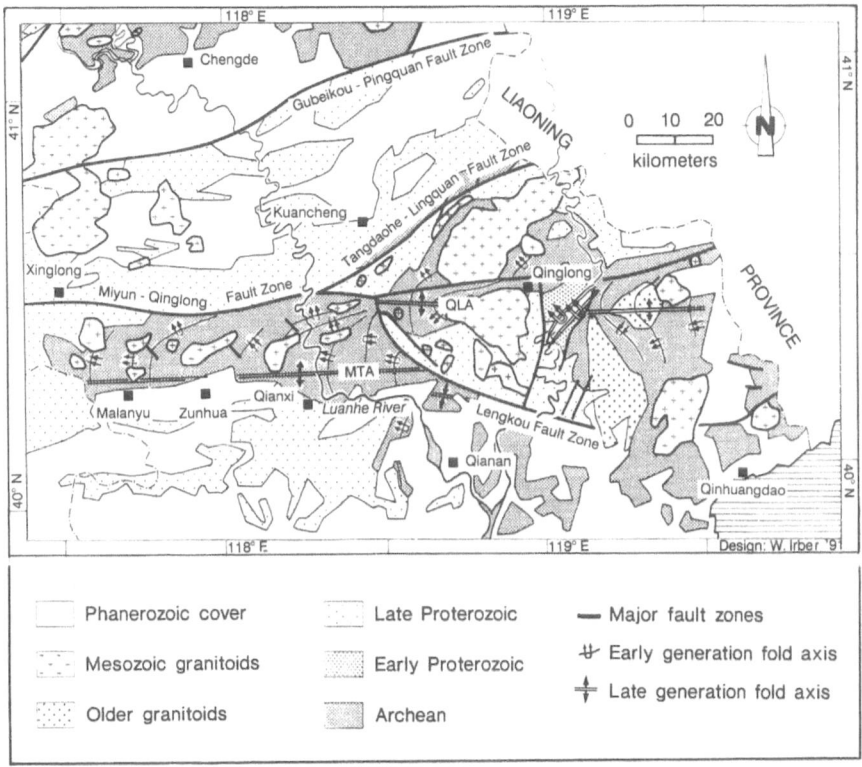

Fig. 2.5. Geologic map of part of eastern Hebei province showing the main structural elements from Sun et al. (1989) discussed in the text

their hinge zones which are represented by ductile shear zones. These facilitated later tectonic movements and magmatic intrusions.

2.3.2 Regional Faults and Lineaments

The most important structural features in northeastern China in terms of gold metallogeny are the faults and fracture zones because they influence the distribution of both magmatic intrusions and hydrothermal circulation. Figure 2.6 shows the most important fault zones in eastern Hebei province and adjacent areas based on geologic and geophysical surveys, super-imposed on a lineament interpretation map derived from 1:250 000 scale LANDSAT images. The rose diagram in the inset shows the dominant NE–SW direction of lineaments and a significant component with N–S and lesser component with E–W trends. These lineaments coincide with known faults and joint patterns measured on the ground, and the mineralized

41

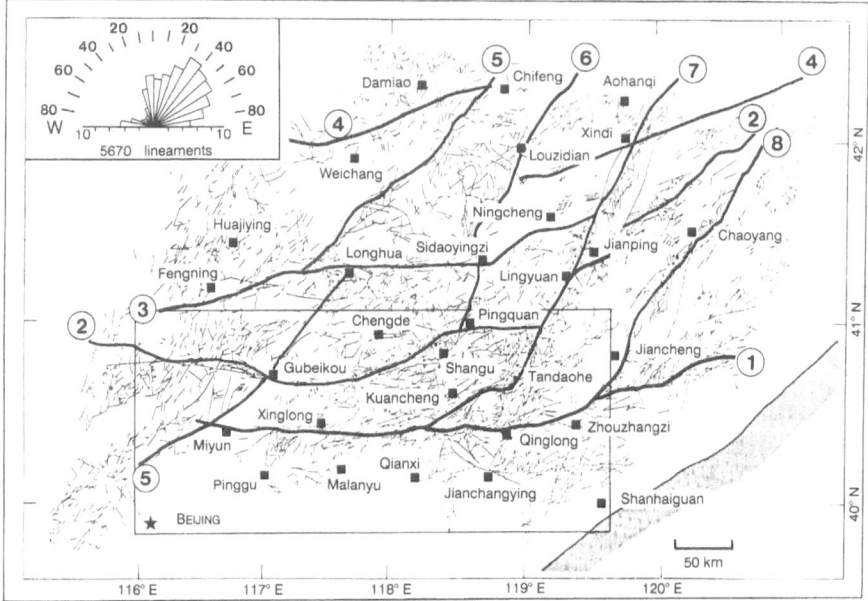

Fig. 2.6. Map showing the regional distribution of lineaments from LANDSAT images of eastern Hebei province and adjacent parts of Liaoning province and southern Inner Mongolia. The numbered fault zones (*heavy lines*) are described in the text. The *boxed area* corresponds to the geologic map of Fig. 1.3

structures in the mining districts also show the dominant NE–SW orientation, as discussed in Chapter 3.

2.3.3 Structural Features of the Major Fault Zones

The area of eastern Hebei province is dominated by NE–SW, N–S, and E–W-trending faults. Multiple faults of a given direction tend to form at regular spacing so that the intersection of the three directions produces rhombic structural blocks. The fault zones are of primary importance in controlling the distribution of gold mineralization both in terms of the location of mining districts and of the associated magmatic intrusions. On a scale of single mining districts the mineralization forms along secondary faults and not directly along the major fault zones.

In the following sections the most important deep fault zones are described in terms of their relative age, orientation, and sense of movement. The fault zones are divided into three groups according to their dominant orientation, namely E–W, NE–SW, and NW–SE, and described in sequence from older to younger. The number given before the name of each fault zone refers to the corresponding number on the map of Fig. 2.6.

2.3.3.1 E–W Fault Zones

The E–W fault zones are developed particularly well in the area south of Longhua. From south to north four major fault zones are recognized:

1. Miyun-Qinglong Fault Zone

This fault zone has a length of more than 300 km. If the secondary faults on either side of the main zone are included, the total width is 10 km. The angle of dip varies, but is generally steep and to the south. The zone is cut by some faults of NE, NNE, and N–S trend, showing that it is relatively early. The geologic nature of the fault zone changes along its length. Most parts of the fault show a reverse sense of movement. Where it separates the Archean gneisses from the Middle to Late Proterozoic rocks, the fault zone shows abundant evidence of ductile deformation including mylonite and associated recumbent folds. The fault zone splays and locally pinches out in the eastern part of the area due to the influence of NNE-trending structures. The overall trend also changes from E–W to ENE in this area.

Most of the Archean exposures in eastern Hebei province are found south of the Miyun-Qinglong fault zone, along the Malanyu-Taipingzhai Anticlinorium (MTA in Fig. 2.5). Numerous intrusive bodies and associated gold mineralizations are distributed along the fault zone.

2. Gubeikou-Pingquan Fault Zone

This zone consists of groups of medium scale faults with E–W and ENE–WSW direction. The faults cut all units from Archean gneisses to Jurassic volcanics. The fault zone is cut by later faults of NW, NNE, and N–S trend, showing that it is relatively early. The western part of the fault zone strikes E–W and dips to the south. The middle part of the zone consists of a series of roughly parallel minor faults and fractures with a NE–SW trend which merge locally with major NE–SW structures. The eastern part of the zone is offset from the other parts by a younger NE-trending fault zone (number 7, Fig. 2.6). In this area the zone consists of an en-echelon series of NE-trending parallel faults. In the Lingyuan district, the fault zone shows reverse movement with the Archean Jianping Group thrust over the Middle and Late Proterozoic rocks.

A marked difference in rock types occurs across the Gubeikou-Pingquan fault zone, which led many geologists to consider that the zone marks the contact between the Inner Mongolian Axis and the Yanshan Fold Belt. Intense vertical movements of the Inner Mongolian Axis relative to the Yanshan Fold Belt during the Yanshan Orogeny led to the intrusion of a zone of mafic-ultramafic rocks along the fault zone at deep crustal levels. Some parts of this zone are related to gold mineralization, notably at Gubeikou, Pingquan, and in the east at Jianping.

3. Fengning-Ningcheng Fault Zone

This fault zone (referred to as the Fengning-Longhua zone in Ren et al. 1987) consists of a major fault and multiple small E–W-trending faults. In many places the zone is cut by NNE- and NW-trending faults. The eastern part of the zone terminates on an intensive major NNE-trending fault zone (number 7, Fig. 2.6). The dip angle of most faults in this zone is near vertical; east of Longhua the faults dip to the north and west of Longhua they dip to the south. The fault zone is most prominently developed in the Archean Dantazi Group. Southeast of Longhua, for example, a cataclastic zone 30–40 m wide is developed. Horizontal slickensides show that the movement involved strike-slip components. According to Ren et al. (1987), the main period of movement of the Fengning-Longhua fault zone was in the Proterozoic.

Many granitic and granodioritic intrusive bodies, most of Yanshanian age, are distributed along both sides of the fault zone. Gold mineralization is particularly abundant at the intersection of this zone with NE-trending fault zones, as especially evident in the Longhua district.

4. Kangbao-Chifeng Fault Zone

The Kangbao-Chifeng zone forms the northern boundary of the Inner Mongolian Axis. The zone does not crop out clearly due to widespread Mesozoic cover, and is not well expressed on satellite imagery. Geologic evidence for a major fault lies in the lithologic contrast across the zone. To the north, Late Paleozoic strata are common, while in the south the Paleozoic is absent and Archean gneisses are widespread. Furthermore, Tertiary basalt and Paleozoic ultramafic intrusions occur along the fault zone. The nature of the fault zone is compressive (reverse movement). The zone is offset by NNE- and NNW-trending faults. Gold mineralization related to the zone occurs mainly where it intersects NNE-trending faults.

2.3.3.2 NE–SW Fault Zones

The NE–SW-directed faults are the most abundant in the area. These are generally younger than the E–W-trending fault zones described above and locally offset them. The most important of the NE–SW-trending fault zones are described below. The numbers refer again to Fig. 2.6.

5. Miyun-Longhua-Mijiayingzi Fault Zone

The fault zone is divided into two separate sections at Longhua (Fig. 2.6). The northern section shows reverse and strike-slip movement where it cuts Jurassic strata. The southern section cuts two major E–W-trending fault zones. The fault zone is interpreted to have formed during the Late Yanshan Orogeny because isotopically dated granites of that age occur along the fault and because it clearly cuts across the earlier E–W-directed fault zones. Gold mineralization is closely related to the fault zone where it intersects E–W zones such as at Longhua, Miyun, and Gubeikou.

6. Pingquan-Balihan-Hongshan Fault Zone

This fault zone can be divided into a southern and a northern section at its intersection with the E–W-trending Fengning-Ningcheng fault zone (number 3). The southern section is a Late Mesozoic feature made up of a dense zone of small faults. The northern section is poorly expressed but both sides of the zone show different rock types, with Archean gneisses and mafic granitoids on the west and Jurassic volcano-sedimentary rocks and Quaternary sediments on the east. The favorable locations for gold mineralization along the fault zone are at points of intersection with E–W faults.

7. Tangdaohe-Lingyuan Fault Zone

This is the largest of the NE-trending fault zones in the region. The southern end of the zone is deflected around the 200-Ma-old Dushan Granite and therefore interpreted to be Yanshanian in age. West of the Dushan Granite the zone branches and connects with the E–W-trending Miyun-Qinglong fault zone (number 1). The major gold mineralization along this fault zone occurs in areas of intersection with E–W-trending faults. Thus the Dushan area is at the intersection with the E–W-trending Miyun-Qinglong fault zone (number 1) and the Jianping district is at the intersection with the E–W-trending Gubeikou-Pingquan fault zone (number 2).

8. Jianchang-Chaoyang Fault Zone

This is a wide zone consisting of densely spaced small faults with good expression on satellite images. The fault zone occurs in Proterozoic, Paleozoic, and Mesozoic rocks. In the southern part of the zone reverse movement thrust Archean rocks over Cambrian and Ordovician strata. The intrusion along the fault zone of a pluton which cuts an Early Yanshanian granite suggests that the zone is Late Yanshanian in age.

2.3.3.3 NW–SE Fault Zones

In addition to the above-mentioned major E–W- and NE–SW-trending fault zones the satellite imagery shows some zones of NW–SE trend. Most of these are small in scale and discontinuous. The largest of the NW–SE trending zones are the Lengkou Fault Zone (or Jianchangying-Shangying Zone), Tangdaohe-Kalaqin Fault Zone, and the Qianxi-Longhua Fault Zone. All of these cut the E–W- and NNE–SSW-trending fault zones, and are therefore younger. The Lengkou Fault Zone is one of the largest zones, with a length of 70 km. The faults within this zone show oblique-slip movement with intense development of mylonite and horizontal slickensides.

2.4 Yanshanian Granites

The Yanshanian Orogeny was a period of major magmatic and tectonic activity which began in the Early Jurassic and affected most of the continental margin of eastern Asia. This was also the most important phase of metallogeny in eastern Asia. The Yanshanian events caused widespread "reactivation" of the Sino-Korean platform, and also of the Yangtze platform to the south. This section summarizes the nature of Yanshanian magmatism, with emphasis on the granitic intrusions. The role of the Yanshanian granites in the formation of the gold deposits of eastern Hebei province is touched on briefly here, and discussed more fully in Chapter 4.4.

It should be noted that the magmatic activity in some parts of eastern Asia continued through the Late Mesozoic and into the Cenozoic era, well past the time period formally assigned to the "Yanshanian" Orogeny (i.e., Early Jurassic to Late Cretaceous). For convenience in this text the term Yanshanian will be used in a broad sense for all Jurassic through Tertiary magmatic events in eastern Asia. In China, as will be described below, most granitic magmatism ceased around 100 Ma and the term Yanshanian is therefore appropriate for the granites in a strict sense.

2.4.1 Regional Characteristics and Plate Tectonic Setting

The magmatic belt of Mesozoic and Tertiary volcanic and plutonic rocks in eastern Asia extends the entire length of the Eurasian continent from the Bering Strait to Indochina and is up to 3000 km in width. Figure 2.7 shows the distribution of only the Late Mesozoic (Jurassic-Cretaceous) plutonic and volcanic rocks in this area according to the 1:8 000 000 scale *Tectonic Map of Asia* (Li et al. 1982).

This magmatic belt forms an important part of the Mesozoic-Cenozoic circum-Pacific magmatic province and, like its counterparts in North and South America, the magmatic activity is intimately related to subduction of oceanic lithosphere. It is therefore useful to review what is known about the Mesozoic plate tectonic evolution of the western Pacific margin. The Mesozoic configuration of plate boundaries and relative motion vectors of the plates in the western Pacific are not well known, especially for the period prior to 150 Ma, which is the age of the oldest magnetic lineations on the Pacific seafloor northeast of Japan (Uyeda and Miyashiro 1974). The Mesozoic configuration of the east Asian continental margin is also poorly known, but most authors consider that the Japanese islands were attached to the continental margin throughout the Mesozoic (Uyeda and Miyashiro 1974; Takahashi 1983; Maruyama et al. 1989). Based on the plate tectonic syntheses of Uyeda and Miyashiro (1974), Dickinson (1979), Takahashi (1983), Maruyama et al. (1989), and Wiley et al. (1990), the following

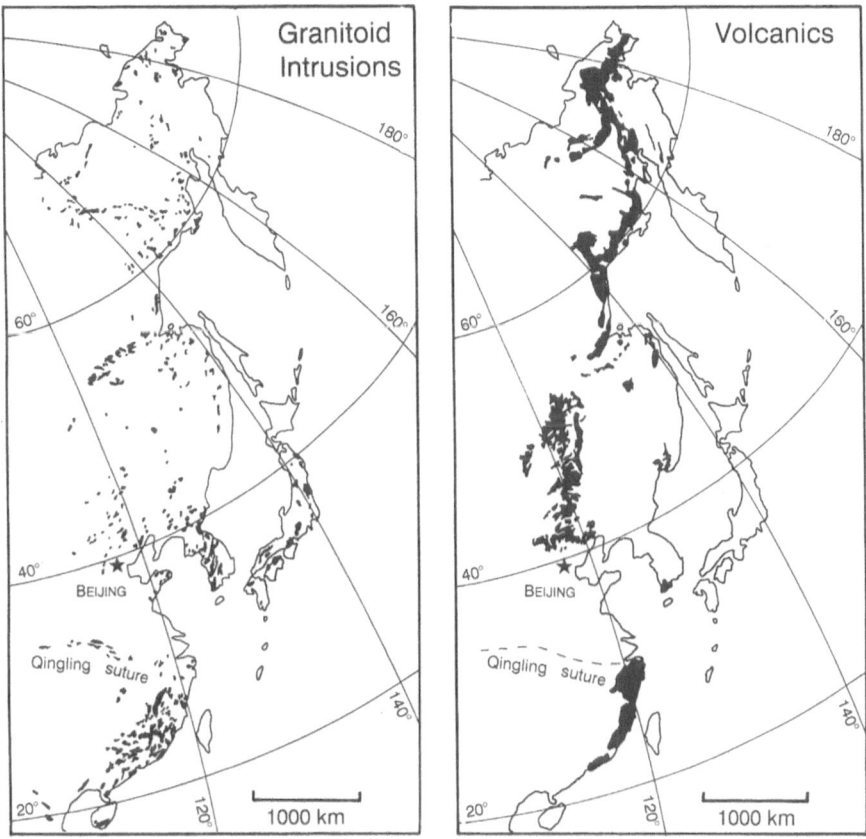

Fig. 2.7. The distribution of Late Mesozoic granitoid intrusions and acid-intermediate volcanic rocks in the eastern margin of Asia, compiled from Li et al. (1982) 1:8 000 000 tectonic map of Asia

important stages in the evolution of the western Pacific margin can be proposed. The age assignments are relatively uncertain and subject to revision:

1. Early Jurassic-Early Cretaceous: subduction of the Izanagi plate under Eurasia along the Kurile and Ryuku arcs. Motion of the Izanagi plate was NNW. Rapid subduction of cold oceanic lithosphere causes high P/low T metamorphism in Taiwan, northeastern Siberia, and north-western Japan.
2. Middle Cretaceous: descent of the Izanagi-Pacific ridge beneath the Japanese islands. Shallow subduction of hot oceanic lithosphere causes extremely broad belt of magmatism on the Eurasian continent.
3. Late Cretaceous to Eocene: subduction of the Pacific plate under Eurasia along the Kurile, Japan, and Ryuku arcs. Motion of the Pacific plate

47

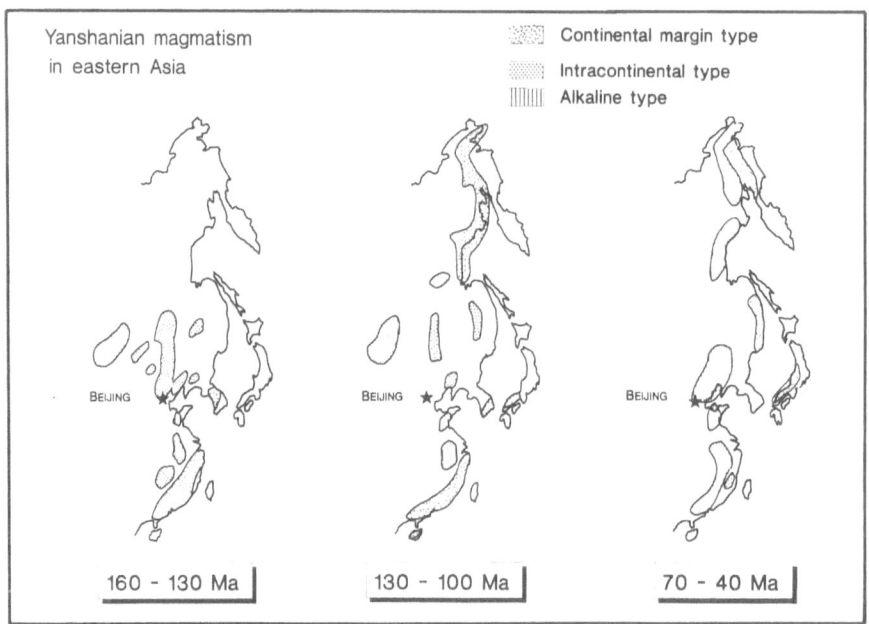

Fig. 2.8. Sketch maps showing the evolution in time, location, and composition of Yanshanian (Mesozoic-Cenozoic) magmatism in eastern Asia. (After Takahashi 1983)

NNW (parallel to the Emperor seamount chain). Incipient opening of the Sea of Japan (?). In the northwestern Pacific, collision of the Okhotsk block (Kamchatka peninsula) with the Eurasian continent.

4. Eocene to present: rotation of Pacific plate motion from NNW to WNW (parallel to the Hawaii seamount chain). Subduction begins in northwest Japan. Formation of the Marianas, Izu-Bonin island arcs and trench systems. Opening of the eastern Asian marginal seas (Sea of Japan, South China Sea, Philippine Sea).

Figure 2.8 illustrates an interpretation of the Mesozoic and Cenozoic evolution of the eastern Asia magmatic belts based on Takahashi (1983). The division of the magmatic rocks into continental margin, intracontinental, collisional and alkaline subtypes is based primarily on the geologic setting and on the whole-rock concentrations of K_2O, Na_2O, and the K_2O/Na_2O ratio.

The sketch maps of Fig. 2.8 show that, in general, the main focus of subduction-related magmatism (the continental margin and island arc types) shifted eastward with time. According to the information available to Takahashi (1983), the main period of igneous activity in northeastern China was Late Jurassic to Early Cretaceous (about 160–100 Ma). After about

100 Ma subduction-related magmatism was confined to southern China, Japan, and the eastern coasts of Korea and Siberia. The Tertiary magmatism in China was alkaline in nature, related to extensional tectonics, whereas subduction-related calc-alkaline magmatism continued in Japan, Sikhote-Alin, and the island arcs around the Philippine Sea.

2.4.2 Yanshanian Magmatism in Eastern China

The Yanshanian magmatic rocks in eastern China clearly form a northern and a southern grouping in Fig. 2.7 with a third minor concentration extending westward along the Qingling suture in central China. Wu (1985) referred to these groups as the northern, southern, and central petrologic/ metallogenetic provinces, respectively. According to his review, the Southern Province contains calc-alkaline volcanic rocks of the andesite-dacite-rhyolite series. Related granitoids are granodiorite, biotite granite, and peraluminous granites. Ore deposits of W, Sn, REE, Nb, Ta, U, and Be are present. The rocks of the Northern Province are calc-alkaline basalt-andesite-rhyolite series volcanics and diorite, granodiorite, and granite intrusives. Related ore deposits contain Mo, Cu, Pb, Zn, Au, and Ag. The Central Province rocks have a different tectonic and lithologic affinity from the northern and southern provinces. They consist of alkaline (K-rich) shoshonitic basalt-trachyandesite-trachyte series. Granitoids are monzodiorite, monzonite, and syenite. Related ore deposits contain Cu, Fe, and Mo. The nature of Yanshanian granitic magmatism in the southern and northern provinces is described below. The central province is not discussed further.

2.4.2.1 Southern Province

The best-studied area of Yanshanian granite intrusions and related metallogenesis is in southeastern China, in the "Southern Province" of Wu (1985). The nature of the magmatism is briefly described below. Several comprehensive reviews of the data exist (Xu et al. 1982; Wang et al. 1983; Wu 1985). Most studies of these rocks have identified two groups of granitoids on the basis of chemical and petrographic criteria. The names of the two groups may vary according to the author but their characteristics are essentially the same, and these are summarized in Table 2.4.

According to a summary of Yanshanian granites in southern China by Ishihara (1984), the group II intrusions (his "ilmenite series") are Early Yanshanian in age (i.e., Jurassic), and they occur in Paleozoic sedimentary rocks in a Caledonian fold zone. Group I plutons (his "magnetite series") are of Late Yanshanian (Cretaceous) age, and they occur in the coastal Fujian volcanic belt, where they intruded during a tensional stress regime. Xu et al. (1982) confirmed that the group I intrusions tend to form near the

Table 2.4. Division of Yanshanian granitoids in southeastern China

Criterion	Group I	Group II
K_2O/Na_2O	0.7–1.7	0.8–2.5
Element enrichments	Cl, Sr, Ba	F, Rb, Li
REE pattern	Inclined, LREE > HREE	Flat
Eu anomaly	None or weak negative	Strong negative
$^{87}Sr/^{86}Sr$ initial	<0.709	>0.710
$\delta^{18}O$ (whole rock)	+5.5 to +10.5	+9.2 to +13
Accessory minerals[a]	Mt, Sph, Zr	Ilm, Mt, Mon, Zr

References: Xu et al. (1982); Wang et al. (1983); Wu (1985); Ishihara (1984).
[a] Ilm, ilmenite; Mt, magnetite; Mon, monazite; Sph, sphene; Zr, zircon.

Pacific margin along deep faults and the group II intrusions form farther inland.

Wu (1985) summarized Rb-Sr isochron age data from a total of 59 granitoids and 16 volcanic units of eastern China as a whole. From the total of 75 age dates, 89% are older than 100 Ma, and 51% fall in the range 200–125 Ma. The intrusions reported by Wu (1985) can be divided into the Group I and Group II-type intrusions of Table 2.4 based on their initial $^{87}Sr/^{86}Sr$ ratios. No systematic difference in age between the two groups is apparent from Wu's data.

2.4.2.2 Northern Province and Korea

Unfortunately, there is not nearly as much information on granites in northeastern China as in the south, and no comprehensive summary of the Yanshanian magmatism in this region has been published as far as the authors are aware. The nearest area which has been well documented is southern Korea (Ishihara et al. 1981; Kim and Lee 1983; Ishihara 1984). In this area, the age and compositional character of the Yanshanian granitoids are quite similar to those described above from southeastern China, except that the ages of the granites in general are younger in southern Korea than in China.

In South Korea, two groups of Yanshanian granitoids are distinguished, the earlier Daebo Group and the later Bulgugsa Group. According to the summary by Ishihara (1984), the Daebo Group is dominantly of Jurassic age (many ages cluster around 150–170 Ma) and it consists mostly of I-type, ilmenite series tonalitic to granodiorite and granite plutons. The younger Bulgugsa Group shows mostly Cretaceous ages (ages cluster around 60–90 Ma) and it consists of high-level granodiorite and granite intrusions of magnetite-series type. The two Yanshanian granitoid groups occur in

NE–SW-trending belts. The belt of Cretaceous Bulgugsa Group intrusions occurs on the southeastern side of the Korean Peninsula nearer to the Pacific plate margin than the Jurassic Daebo Group intrusions. The Bulgugsa intrusions are closely related to Mo and W mineralization (Ishihara 1984).

2.4.2.3 Summary

Ishihara (1984) discussed a unifying model of Yanshanian magmatism in eastern Asia which relates metallogeny, time, spatial distribution, and the nature of the rocks. His model refers to "paired ilmenite-series and magnetite-series plutonic belts" which form at active continental margins. According to the paired plutonic belt model, an inner belt of ilmenite-series mesozonal plutons without significant volcanism forms near the plate margin. The source of the granites may be mixed igneous and metasedimentary materials with a strong contribution from the continental crust (low oxygen fugacity, high $\delta^{18}O$, high initial Sr isotopic ratio). Associated mineralization of Sn and W derive their metals from the continental crust. Farther from the active continental margin, in a regime of high heat flow and extensional stress, a belt of high-level magnetite-series plutons with coeval volcanic rocks forms along structurally weak zones. The associated mineralization is dominated by Mo. The source of these magmas is likewise mixed but the continental contribution is less important, or restricted to lower-crustal material, and sedimentary input is minimal (high oxygen fugacity, low $\delta^{18}O$, low initial Sr isotopic ratios).

The ideal pattern of coeval ilmenite-series plutonic belts with Sn and W mineralization near the plate margin and magnetite-series plutonic belts with Mo mineralizations on the continental side is complicated in eastern Asia by the eastward migration of the subduction zone with time and opening of the marginal seas. This migration of the plate margin caused overlapping or superposition of younger magnetite-series belts onto older ilmenite series belts. In southeastern Asia, where the geometry of Mesozoic subduction zones was complex and unstable, no clear pattern of plutonic and metallogenetic belts can be seen (Hutchinson and Taylor 1978).

More data on the Yanshanian magmatism needs to be collected in northeastern China before the nature of the rocks and their metallogenetic importance can be fully assessed. By analogy to the well-studied parts of the Yanshanian magmatic province in Asia, it may be expected that systematic patterns of age and compositional distribution exist in northeastern China, and that these patterns will be important to an understanding of metallogeny in that region. Recently, several Chinese geologists have emphasized the probable connection of Yanshanian granites and gold deposits in northeastern China (M.Z. Yang 1988, Liu 1989, Zhou 1989; see also the discussion in Chap. 4.4).

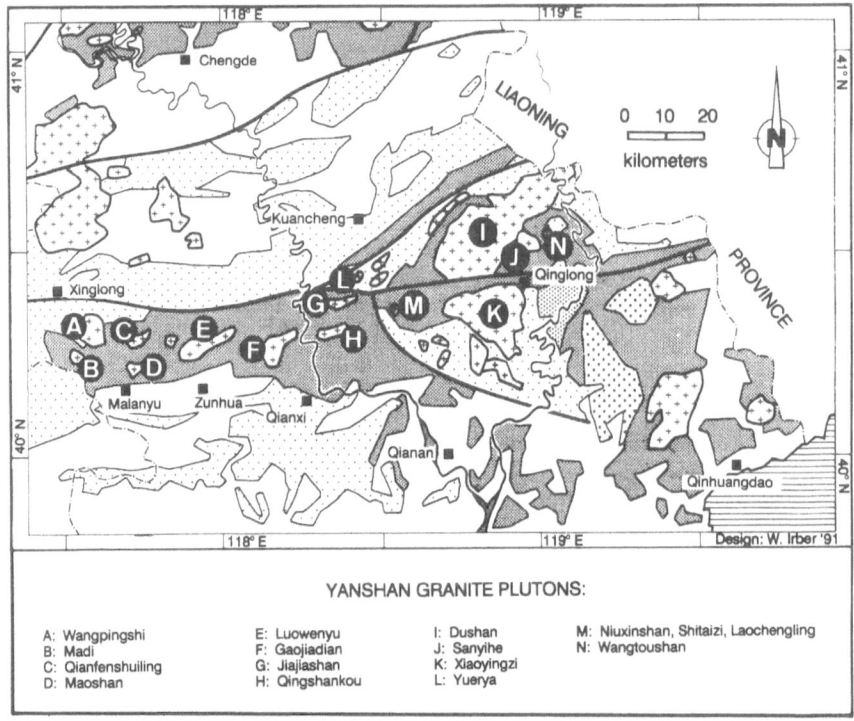

YANSHAN GRANITE PLUTONS:

A: Wangpingshi	E: Luowenyu	I: Dushan	M: Niuxinshan, Shitaizi, Laochengling
B: Madi	F: Gaojiadian	J: Sanyihe	N: Wangtoushan
C: Qianfenshuiling	G: Jiajiashan	K: Xiaoyingzi	
D: Maoshan	H: Qingshankou	L: Yuerya	

Fig. 2.9. Geologic map of part of eastern Hebei province showing the Yanshanian plutons which were sampled in this investigation and discussed in the text. For explanation of *symbols*, see Fig. 1.3

2.4.3 Yanshanian Granites in Eastern Hebei Province

Despite the apparent importance of the Yanshanian granites to the gold deposits of eastern Hebei province, no regional study of the Yanshanian granites in that area has yet been published. In the course of our joint investigations, a total of 16 plutons in eastern Hebei province were sampled in a first attempt to systematically characterize the Yanshanian granites in this gold province. The distribution of plutons sampled is shown in Fig. 2.9, and a table of their average compositions is given in Appendix 1. Age data are available for only eight of the sixteen plutons, but according to the available data (range about 200–160 Ma), the first Yanshanian stage appears to have been important for plutonism in eastern Hebei province. In general the granites form small (outcrop area often less than 10 km^2) discordant plutons which are elongate in the NNE-SSW and ENE-WSW directions, parallel with major fault zones and regional lineaments. The granites are often located at the intersection of fracture zones, and their textures and

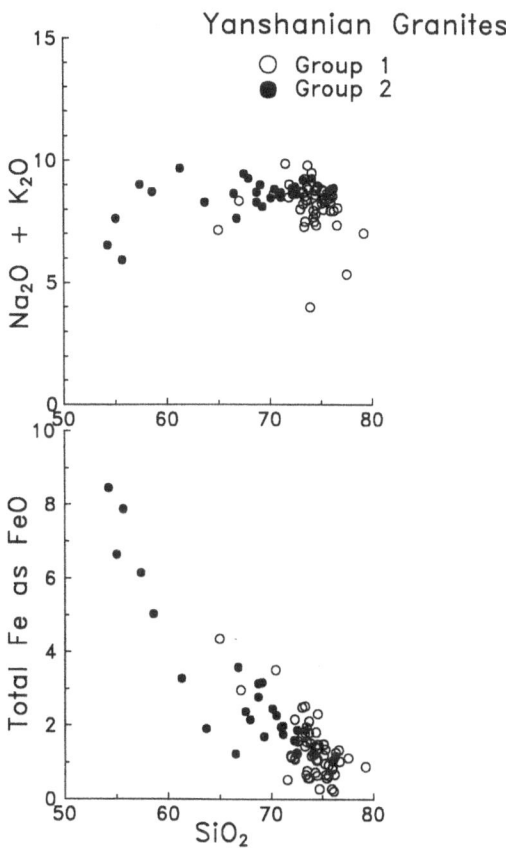

Fig. 2.10. Harker diagrams showing the compositions of 16 Yanshanian granitic plutons from eastern Hebei province. The division of the plutons into groups 1 and 2 is empirical according to composition, as discussed in the text

form suggest that they were emplaced at upper crustal levels. Volcanic rocks are rare in this area, although igneous dikes are common.

Two compositional groups of granites can be distinguished based on our limited analyses. The first group includes all but three of the plutons, and could therefore be considered the "normal" Yanshanian plutons in this area. These rocks classify as granodiorite and true granite in the Streckeisen classification. There is considerable variation in composition among the plutons but the variation is smooth and there is no justification for subdividing them based on the present information. The granites are generally biotite-bearing and have $SiO_2 > 70$ wt% and total Fe (as FeO) up to 3 wt%. Some plutons of this group have mildly peraluminous compositions but most are slightly metaluminous or just saturated in Al. The second group of granitic plutons is considerably more mafic than the former group, and the

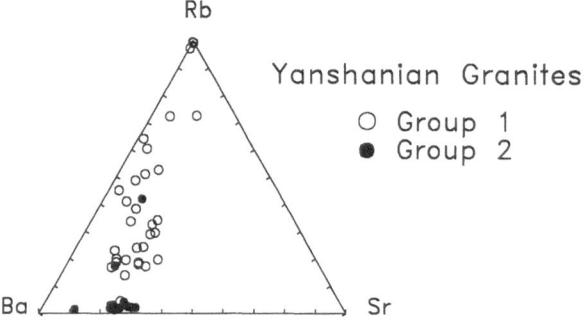

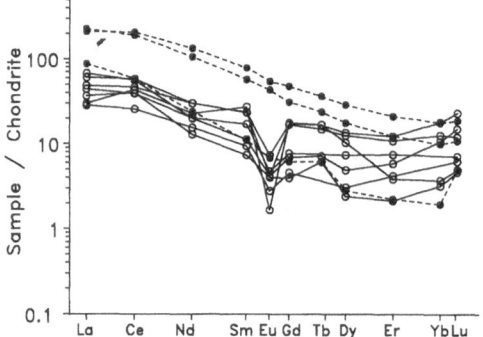

Fig. 2.11. Rb-Sr-Ba diagram and REE distribution pattern of selected Yanshanian granitic plutons from eastern Hebei province. Groups 1 and 2 plutons are the same as in Fig. 2.10

plutons plot as granodiorite to quartz diorite in the Streckeisen classification. They are hornblende and biotite-bearing, have SiO_2 concentrations generally under 70 wt%, and total Fe (as FeO) concentrations from about 2 to 8 wt%. They are mildly to strongly metaluminous. Figure 2.10 shows the major element distinction between the two groups in Harker diagrams of total alkalis and iron. Figure 2.11 shows the distribution of Rb-Sr-Ba and REE in the two groups of plutons. In both diagrams, the group 2 plutons show markedly less differentiated character.

Based on their major element chemistry (AFM, Fe/Mg vs SiO_2), all plutons belong to the calc-alkaline series (Wilson 1989). According to their whole-rock ratios of $Fe_2O_3/(Fe_2O_3 + FeO)$, the granites belong to both the ilmenite series and the magnetite series of Ishihara (1984). Of the 16 plutons, seven clearly classify as magnetite-type, five clearly classify as ilmenite type and four are transitional in character based on Ishihara's (1984) criterion. Both the group 1 and 2 plutons contain members of both the ilmenite and magnetite series. Initial $^{87}Sr/^{86}Sr$ isotopic data are presently available for only the Niuxinshan and Sanyihe Granites (see Chap. 3). Both have low initial $^{87}Sr/^{86}Sr$ ratios of about 0.704, which suggest a derivation from the relatively primitive Archean lower crust. M.Z. Yang (1988) also attributes the Yanshanian granites within Precambrian uplifts in north-eastern China to anatexis of the lower crust.

The geologic relationship, and especially the time relationship between these two granite groups are not presently known, nor is their relative importance to gold metallogeny understood (see Chap. 4.4). The mafic plutons of group 2 form the largest intrusions in the area and in general these are not associated with gold occurrences. The group 2 plutons may be older than those of the first group; for example, the Dushan Granite (200 Ma) is clearly older than the dated plutons of group 1, which span ages of 160 to 190 Ma, but further studies are needed to clarify the age and genetic relationship between these two plutonic groups.

3 Description of Selected Gold Deposits

This chapter describes the geology, mineralogy, and geochemistry of selected gold deposits in five mining districts of eastern Hebei province whose location is shown in Fig. 3.1. There is commonly more than one gold deposit in each district, and the most important deposits are listed here for reference, the numbers corresponding to the relevant chapter where the district is discussed. Only the deposits in italics are described in detail:

3.1 Niuxinshan District: *Niuxinshan*, Huajian deposit.
3.2 Sanjia District: *Sanjia, Wangtoushan, Xinglonggou* deposits.
3.3 Yuerya District: *Yuerya* deposit.
3.4 Jinchangyu District: *Jinchangyu* deposit.
3.5 Banbishan District: *Banbishan*, Zhangzhangzi deposits.

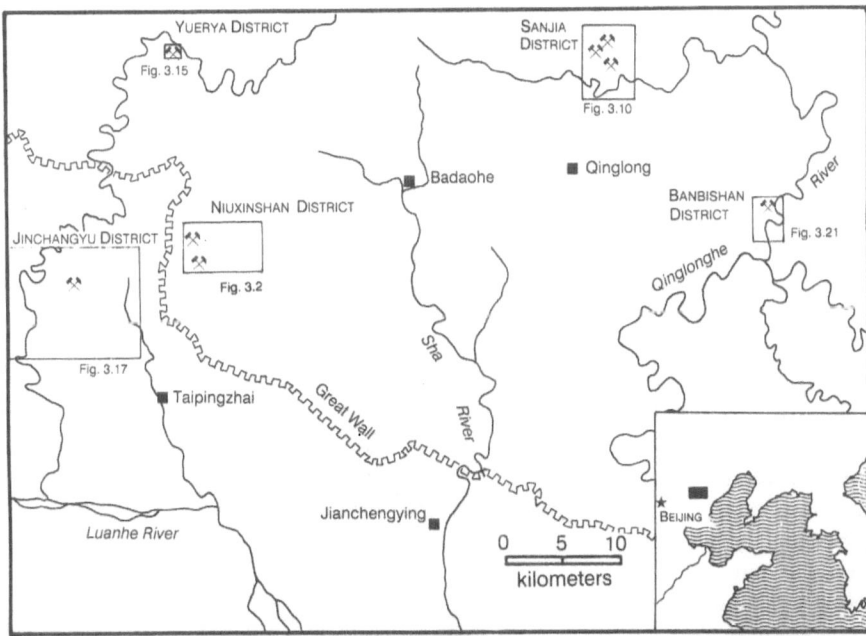

Fig. 3.1. Map of part of eastern Hebei province showing the location of the mining districts described in the text. *Inset* shows area of main map

The mining districts listed above include typical representatives of the various gold deposit types present in eastern Hebei province as discussed in Chapter 1.3 (with the exception of placer deposits). Deposits in Archean amphibolite to granulite-facies supracrustal rocks are represented by the Niuxinshan, Sanjia, and Jinchangyu districts. In fact, the Jinchangyu deposit is the type deposit of the Archean-hosted deposits according to Zhu's (1989) classification. The Banbishan district is a good example of metamorphic-hosted gold deposits in Early Proterozoic low- to medium-grade metagray-wackes. Granite-hosted gold deposits are represented by the Yuerya district, the type deposit of this category (Zhu 1989), and by parts of the Niuxinshan and Sanjia districts.

The reader will note that not all deposits are described below in the same detail. This is because the amount and the sources of information available about the deposits varies. Much of the description of the Niuxinshan and Sanjia districts is based on new data collected during the 1987–1989 project investigations. The descriptions of the Jinchangyu, Yuerya, and Banbishan deposits are based on published studies and unpublished reports of the Ministry of Metallurgical Industry.

3.1 Niuxinshan District

The gold mining district of Niuxinshan is located about 200 km northeast of Beijing in the county of Kuangcheng, Hebei province. The district is divided into four sections: Niuxinshan, Huajian, Maweigou, and Yümoling, as shown on the map of Fig. 3.2. Gold is produced from only the Niuxinshan and Huajian deposits, and the latter was nearly mined out as of 1989. The Maweigou and Yümoling sections of the district are in the exploration stage. In the following discussion, emphasis is placed on the Niuxinshan deposit. The gold deposits at Niuxinshan have been known since the Qing Dynasty (1644–1911). The mines are operated by the Kuancheng County govern-ment. The ore is treated at a central concentration plant in the village of Huajian.

The Niuxinshan Deposit
The Niuxinshan deposit is the largest in the district. Figure 3.3 shows a simplified geologic map of the deposit. Gold is won from more than 20 quartz-sulfide veins which strike NE–SW and dip moderately to the NW. Mineralization occurs mainly in mafic metamorphic rocks but also in granite. The average gold grade is 15–20 g/t. The mine workings are entirely underground.

The Huajian Deposit
The geologic setting of the Huajian deposit is the same as at Niuxinshan but the mineralization at Huajian is entirely in the metamorphic rocks. The ore

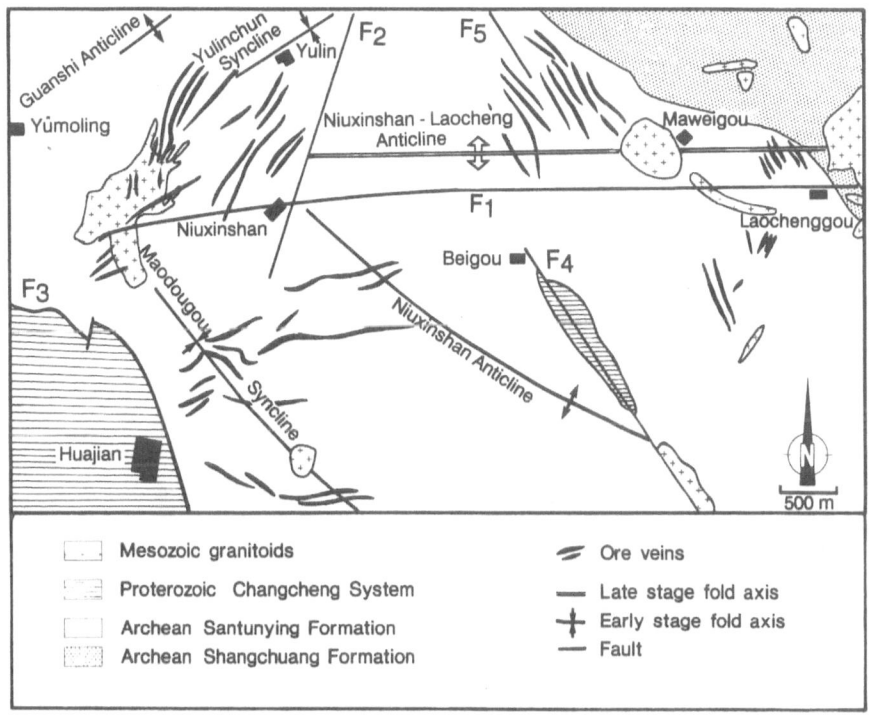

Fig. 3.2. Simplified geologic map of the Niuxinshan mining district

veins strike mostly E–W and dip at a moderate angle to the N (see Fig. 3.2). There are some 35 gold-bearing quartz veins known in the deposit. Gold grade of the ore bodies ranges from 7–14 g/t, but most ore has been mined out. Mining is entirely underground.

3.1.1 Host Rock Lithologies

Archean gneisses, amphibolites, and granulites of the Qianxi Group form the main host rocks in the Niuxinshan district. In the west and southwest of the district, sandstones of the Middle Proterozoic Changcheng system (Dahongyu Formation) are in contact with the Archean rocks along a high-angle reverse fault.

The Proterozoic rocks are resistant to erosion and they form a high crest through the region upon which part of the famous Great Wall of China is built, forming an imposing backdrop to the deposit (see frontispiece).

Igneous rocks in the district are represented by several small granite stocks of Yanshanian age and innumerable dikes of mafic and felsic composition. Most of the dikes postdate the granites and predate the mineralization. Of the granite intrusions, only the Niuxinshan Granite is mineralized.

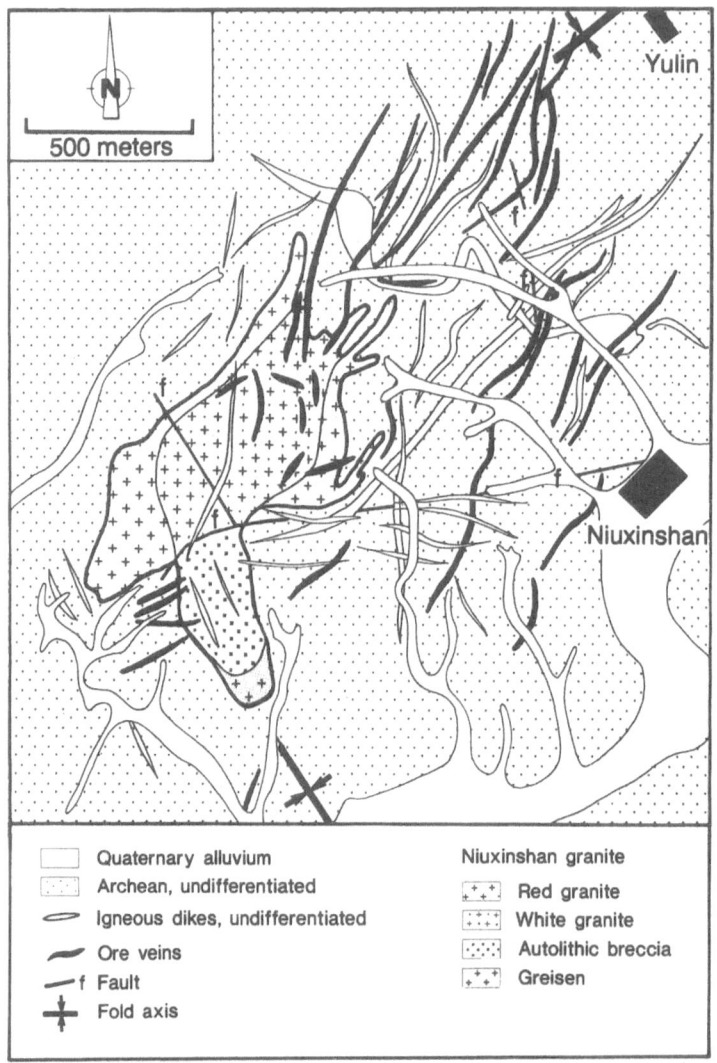

Fig. 3.3. Simplified geologic map of the Niuxinshan deposit

3.1.1.1 Archean Rocks

The Archean rock sequence in the Niuxinshan district comprises the lower two formations of the Qianxi Group. The lowest, Shangchuang Formation, is composed of granulite-facies mafic and intermediate gneisses (hornblende-pyroxene gneisses, pyroxenites, two-pyroxene gneisses). The Shangchuang Formation is exposed only in the northeastern part of the district near Maweigou (Fig. 3.2), and it does not form the host rock for any of the gold deposits. The Santunying Formation structurally overlies the Shangchuang Formation and it forms the host rock of all deposits in the Niuxinshan district. The formation contains a variety of rock types dominated by migmatized amphibolites and hornblende-biotite gneisses of variable mineralogy and texture. The rocks of the Santunying Formation are similar to the rocks of the Shangchuang Formation except for the higher metamorphic grade of the latter.

Retrograde metamorphism in both the amphibolites and the granulites is widespread and expressed as saussuritization of plagioclase and uralitization of pyroxene and hornblende. Migmatization is well developed in both formations and there are few rock outcrops entirely free of leucosomes. Rocks containing more than 50% of leucosomes, referred to as migmatitic granite in the Chinese reports, occur as diffusely bound areas of several tens to hundreds of square meters in extent. These areas are commonly conformable to the lithologic layering and foliation of the amphibolite sequence but may locally transect them.

A minor but important constituent of both the Santunying Formation and the Shangchuang Formation are discontinuous layers or lenses some meters thick of quartz-magnetite amphibolite. These quartz and magnetite-rich rocks grade continuously over a few decimeters from normal amphibolites into massive magnetite quartzite with over 30% by weight of iron, which are a local source of iron ore. At Niuxinshan the magnetite quartzite lenses are concentrated in a NNE-trending zone through the center of the mineralized area. The magnetite-rich rocks are intensely pyritized where they are cut by quartz veins, and this alteration is an important local factor for gold mineralization. The importance of iron-rich rocks for gold mineralization in the Archean-hosted deposits is discussed in detail in Chapter 4.3. Other minor rock types, which occur as lenses or discontinuous layers in the amphibolite and gneiss sequence, include coarse-grained garnet-diopside-plagioclase gneisses and coarse hornblendites or hornblende pyroxenites. The latter rocks may represent former gabbroid and ultramafic dikes.

All of the Archean rocks are strongly foliated and in many outcrops the rocks show isoclinal folds. The structures are described separately in Chapter 3.1.3.

3.1.1.2 Granite and Dikes

Several small stocks of granite occur along or near major NE-SW- and E-W-trending faults and fracture zones in the Niuxinshan district, as seen in Fig. 3.2. The granites are all assigned a Yanshanian age because they are undeformed, discordant to the host rocks, and because the fault zones with which they are associated are thought to have formed during the Yanshan Orogeny. Isotopic dating of the Niuxinshan Granite (see below) is consistent with the Yanshanian age; the other granites have not been dated.

The Niuxinshan Granite is the only one of the granites which has been studied in detail. The granite forms a composite pluton with irregular elliptical shape in map view, elongate to the NE. Two petrographic varieties have been distinguished: an inner, medium- to fine-grained equigranular whitish-gray granite and an outer medium-grained porphyritic red granite. Both the white and red varieties of granite have essentially the same mineral assemblage, with an estimated 5% biotite, 25% quartz, 40% oligoclase, and 30% K-feldspar. Accessory phases include white mica (probably primary), zircon, allanite, and opaque minerals. Field relations suggest that the inner, white granite is the younger. This is supported by Rb-Sr isochron ages of 229 ± 33 Ma for the red granite and 174 ± 14 Ma for the white granite (Ministry of Metallurgical Industry, unpubl. data). The $^{87}Sr/^{86}Sr$ initial ratios measured from the two granite phases are 0.714 ± 0.003 and 0.704 ± 0.004 respectively.

At the southern end of the pluton a large body of breccia occurs which extends into the country rocks. The breccia consists of centimeters to meter-sized angular blocks of granite and/or amphibolite cemented by quartz with associated white mica, fluorite, hematite, and rare sulfide minerals. The breccia body is interpreted to have formed by abrupt release of magmatic volatiles at a relatively shallow level of intrusion.

The chemical compositions of both granite types are quite similar, characterized by 74–76 wt% SiO_2, 7–9% total alkalis (Na_2O/K_2O near 1), 12–13 wt% Al_2O_3, and very low concentrations of the other major element oxides (the sum of TiO_2, MgO, total Fe as FeO, and CaO is less than 1.5 wt%). The granite is metaluminous. The average concentrations of Sr (27 ppm), Ba (224 ppm), and Rb (314 ppm) indicate a moderate degree of differentiation. The chondrite-normalized REE distribution (not shown) is regular and flat, with a moderate negative Eu anomaly (for analytical data see Appendix 1).

A large number of dikes of various composition occur in the Niuxinshan district. Their intrusive relations with the granites and ore veins place important constraints on the timing of mineralization. The earliest dikes consist of diabase and they predate the intrusion of the granite stocks. All other dikes can be observed to cut the granites. The earliest of the post-granite dikes are granitic in composition and they consist of variably porphyritic aplite or rhyolite, or rarely pegmatite. The granitic dikes may be cogenetic with the stocks since they are abundant only in areas where granites occur. Granitic dikes are not found, for example, in the Huajian area, where only one small granite stock is exposed, and are most abundant at Maweigou, where six granite intrusions occur. The granitic dikes are cut by dikes of porphyritic diorite. These dikes are cut in turn by dikes of porphyritic quartz trachyte. The youngest dikes in the district consist of lamprophyre which cut all other dikes and the granites.

Field relations show clearly that the granitic/aplitic dikes are cut by the ore-bearing quartz veins and weakly mineralized. The dioritic dikes may locally be cut by quartz veins, but often the dikes cut the veins; thus their intrusion seems to span the time of vein formation. The quartz trachyte and lamprophyre dikes consistently cut across the ore veins. Sun et al. (1989) reported K-Ar ages of 209 and 166 Ma from dioritic and lamprophyre dikes, respectively, at the Huajian deposit. No other dates from dikes in the Niuxinshan district have been obtained.

3.1.2 Host Rock Structures

The structure of the basement rocks in the Niuxinshan district is influenced by (at least) two main periods of Precambrian folding and by younger faults and fractures zones. The granites and dikes intruded after the folding episodes and show evidence of only brittle deformation.

3.1.2.1 Folds and Foliation

Both early and late-stage folds are well represented in the Niuxinshan district. The late-stage folds are open and upright with E–W-trending axes and little or no development of axial planar foliation. The prominant example of the late-stage folds is the Niuxinshan-Laocheng anticline, which runs through the center of the district (Fig. 3.2). The early folds are tight to isoclinal and generally overturned, with westward-dipping axial planes. They are refolded about the E–W anticline so that their axes trend NE–SW in the northern part of the district and NW–SE in the southern part. The gneissic foliation is axial planar to the early generation of folds and it therefore strikes NW–SE on the southern flank of the late stage anticline, NE–SW on the northern flank and approximately N–S in the hinge zone.

The hinge zones of the folds are preferred sites for later faults and igneous intrusions. Examples shown in Fig. 3.2 are the F1 fault and the granite stocks and dikes intruded along the hinge zone of the Niuxinshan-Laocheng anticline, and the granite along the hinge of the NW-trending Maodougou syncline in the Huajian area.

3.1.2.2 Faults and Fractures

Faults and fractures are important structural elements in the Niuxinshan district and they ultimately control the distribution of the ore-bearing veins. Zhao (1989) studied the structures and related mineralization in the Niuxinshan district in detail and divided the fracture/fault zones into three groups based on their strike direction, namely, NW–SE, N–S, and NE–SW. The largest of the fault zones are marked on the district map (Fig. 3.2) with the numbers F1 through F5. No sequential order is implied by the numbers.

NW–SE-Trending Faults

The NW–SE trending group of fault zones forms part of the regional Lengkou fault zone discussed in Chapter 2.3.3 (see Fig. 2.5). The F3, F4, and F5 faults are parts of this system and all are SW-dipping reverse faults. Fractures parallel to this group are the dominant host structures for dikes and quartz veins in the Maweigou section of the district and to a lesser extent also in the Huajian section. Related vein fillings are massive milky quartz with only minor sulfide minerals and little gold; with a few exceptions they have no economic significance.

E–W-Trending Faults

The main E–W-striking fault zone in the Niuxinshan district is termed the F1 fault. This is a planar strike-slip shear zone which dips 55°–65° to the south. Fractures with the E–W trend are host to andesitic and granitic dikes in the Maweigou and Huajian areas, and at Niuxinshan the late-stage lamprophyre dikes also trend E–W. However, except for some veins at Huajian, the E–W structures are rarely important hosts for ore-bearing quartz veins.

NE–SW-Trending Faults

The NE–SW-trending fracture zones and faults dominate in the Niuxinshan area, and the main phase of mineralization in the Niuxinshan deposit occurs in quartz veins with NE–SW trend. The major NE–SW fault zone in the Niuxinshan district is designated F2 and it is an oblique-slip fault dipping 60°–70° NW which extends more than 10 km. Dikes and ore-bearing quartz veins occupy NE–SW structures in the Niuxinshan area and, to a lesser extent, in the Huajian area. NE–SW trends are not developed in the Maweigou section of the district.

3.1.3 Gold Mineralization

The gold mineralization in the Niuxinshan district is confined to quartz veins and selvages of altered wall rock. The host rocks are the Archean amphibolites and gneisses of the Santunying Formation, local exceptions being the Niuxinshan Granite and some granitic dikes. The ore paragenesis and style of mineralization are essentially the same throughout the district and can be exemplified by the Niuxinshan deposit, on which the following description is based. The only significant differences between the Niuxinshan deposit and the other deposits and prospects are the orientation of the quartz veins (NE–SW at Niuxinshan, NW–SE at Maweigou, E–W and NW–SE at Huajian), and the fact that the granite at Niuxinshan is mineralized whereas other granites in the district are barren.

3.1.3.1 Form of the Ore Bodies

The mineralized quartz veins in the Niuxinshan deposit strike dominantly
NNE–SSW to NE–SW and dip to the NW. Most veins show evidence of
movement and multiple vein-filling, with trains of sheared country rock
fragments within the quartz and striae on the vein walls. The richest mineralization is found in intensely brecciated quartz veins which are cemented
by sulfide minerals. Most of the gold-bearing quartz veins are located in
the Archean metamorphic rocks to the north and east of the Niuxinshan
Granite. More than 100 veins are known, with 20 of them being productive.
The present workings and drill hole data have proved mineralization to a
depth of about 500 m. The veins have an undulatory shape in detail, and
their thickness varies from a few centimeters up to 2.5 m (average 20–
50 cm). It is common to find that the veins occur in groups within relatively
narrow zones. A typical example is the No. 1 vein zone, which contains up
to ten single veins and can be traced for over 1200 m in length and 300 m in
width.

Where the veins cut the Niuxinshan Granite they tend to disperse into thin
splays of quartz veinlets and form a zone of intense greisen-type alteration
with quartz, muscovite, pyrite, and minor fluorite. Several zones are being
worked in the granite but they do not contribute much to the total gold
production of the Niuxinshan deposit even though the grade can reach
several hundred g/t. The breccia zone on the southeastern flank of the
Niuxinshan Granite contains a similar hydrothermal assemblage of quartz,
muscovite, and fluorite, but pyrite is rare and assays show that the zone has
no significant gold mineralization.

3.1.3.2 Macroscopic Description of the Ores

The form and mineralogy of the ore-bearing veins in the Niuxinshan deposit
varies considerably depending on the host rock and on the proximity to the
granite.

Veins in Amphibolite

In amphibolite, the mineralization occurs within decimeter- to meter-
thick quartz veins. Wall rock alteration is present to a width of several
centimeters or more depending on the degree of shearing, but the alteration
zones are not significantly mineralized. Quartz, pyrite, and various base-
metal sulfides dominate the vein assemblage; minor chlorite and sericite
occur around fragments of wall rock. Thin veinlets of carbonate (mostly
dolomite) are common in a cross-cutting relationship to the quartz veins.
Sulfide minerals in the veins occur preferentially at the wall rock selvages,
around inclusions of wall rock within the veins, or as thin seams and nests in
quartz which parallel the walls and represent multiple reopening of the

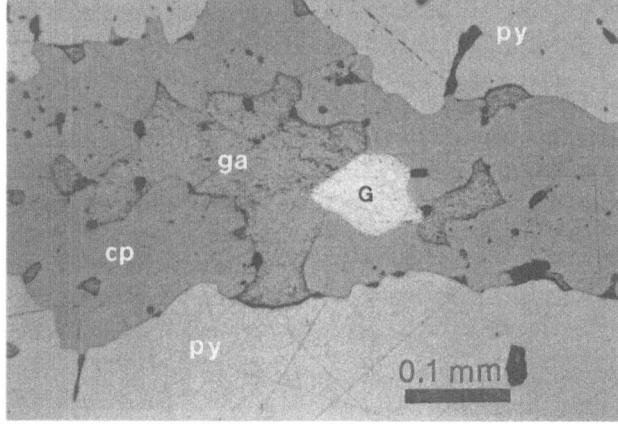

Fig. 3.4. Ore from the Niuxinshan deposit. *Above*: Brecciated quartz vein healed by sulfide minerals. *Below*: Photomicrograph showing the typical occurrence of gold (*G*) with chalcopyrite (*cp*) and galena (*ga*) filling cracks in pyrite (*py*)

veins. The typical sulfide concentration in the veins is 5–10 vol%. In the richest ores the quartz is brecciated and cemented by a network of sulfides which can make up 30% of the vein. Figure 3.4 shows an example of this type of ore.

In the vicinity of the Niuxinshan Granite and/or the granitic dikes the quartz veins bear reddish hydrothermal K-feldspar in seams along the wall rock contact. Fluorite also occurs in the veins or in their alteration envelopes near the granite contact but fluorite is absent elsewhere in the deposit.

Veins in Granite

Within the granite the mineralization occurs in highly fractured, pervasively greisenized zones some decimeters to meters thick with a network of millimeters- to centimeters-thick quartz veinlets. The ore minerals are almost exclusively pyrite with very minor base-metal sulfides. Gangue consists of quartz, muscovite, and fluorite. Gold occurs late in the paragenetic sequence together with rare galena, chalcopyrite, and sphalerite, and fills cracks or grain boundaries within granular pyrite aggregates.

3.1.3.3 Ore Petrography and Paragenesis

The petrographic descriptions given below pertain only to the mineralization within amphibolite wall rocks. The mineralization in the granite is of minor importance, and it was described above.

The mineralized quartz veins within amphibolite in the Niuxinshan deposit contain the following ore minerals: chalcopyrite, galena, native bismuth, native gold, pyrite, scheelite, sphalerite, sulfosalts of the alkinite group, tetradymite, and secondary covellite.

Pyrite

Pyrite is by far the most abundant sulfide mineral. It is found as disseminated euhedral cubes up to several millimeters in size within altered host rocks and as coarse subhedral aggregates in the quartz veins. Pyrite in the wall rocks – in contrast to vein pyrite – shows little or no cataclastic deformation and is rarely associated with other sulfides. The pyrite in quartz veins occurs throughout the paragenetic sequence although most of the pyrite formed early. This early generation of pyrite is almost always fractured and the fractures are filled with galena, sphalerite, chalcopyrite, gold, or gangue quartz and carbonate (Fig. 3.4). A later generation of pyrite can be recognized by its inclusions of galena, chalcopyrite, and gold. A third generation of pyrite occurs in cross-cutting fractures with carbonate. This generation of pyrite is generally not associated with other ore minerals.

Chalcopyrite, Sphalerite, and Galena

Chalcopyrite occurs in both the ore veins and in altered wall rock. In the wall rocks the chalcopyrite forms isolated subhedral grains associated with pyrite. Within the quartz veins most chalcopyrite is found as anhedral fracture-fillings in pyrite, together with galena, sphalerite, and gold. In weathered ore samples, chalcopyrite is locally rimmed by covellite.

Sphalerite occurs mostly in coarse anhedral aggregates intergrown with galena and pyrite in brecciated quartz. It is also found associated with chalcopyrite and galena as fracture-fillings in pyrite. Exsolution blebs and schlieren-like inclusions of chalcopyrite in sphalerite are common.

Galena occurs with sphalerite and pyrite cementing brecciated quartz and as fracture-fillings in pyrite with sphalerite, chalcopyrite, and gold. Galena is the youngest of the base-metal sulfides, and it can be found veining both sphalerite and chalcopyrite, perhaps due to remobilization. Galena is often associated with gold. It frequently contains inclusions of native Bi and of tetradymite.

Native Gold
Gold has been observed in the following forms:

a) enclosed by and intergrown with galena, sphalerite, and/or chalcopyrite in cracks within pyrite; this is the most common form;
b) as rounded inclusions in early (cataclastic) pyrite;
c) as fracture-fillings in quartz together with pyrite.

The grain size of gold is mostly between 0.1 and 0.3 mm. Microanalysis of gold grains by SEM-EDS showed that most of the gold contains 30–35 wt% Ag but some grains, particularly those in quartz, have lower Ag concentrations of 5–15%.

Scheelite
Scheelite occurs mainly within the selvage zones of quartz veins associated with sphalerite and other sulfides. It also occurs as small isolated grains in brecciated and altered amphibolite wall rocks. In the veins the grain size can reach 1 cm. The short-wave fluorescence color of all the scheelite found is typically whitish-blue, which indicates low Mo concentrations.

Tetradymite, Native Bi, and Cu-Pb-Bi Sulfosalts
Tetradymite and native bismuth occur as brightly reflectant laminar inclusions of up to 0.1 mm size in galena. Inclusions in galena with slightly weaker reflectivity than tetradymite proved by electron microprobe analysis to have a composition close to the chemical formula $CuPbBiS_2$, which is akin to the aikinite group of sulfosalt minerals. Tetradymite is the most abundant of all the Bi-bearing minerals and it probably accounts for most of the Bi analyzed in the ores (Table 3.1).

3.1.3.4 Chemical Composition

Table 3.1 shows the compositions of nine grab samples of well-mineralized ore taken from veins in amphibolite host rocks, and one sample (6134) taken from a strongly mineralized greisen zone in the Niuxinshan Granite. The purpose of the table is to show typical element associations and give a rough idea of the concentrations; the values are not necessarily representative for the bulk deposit.

The ores in amphibolite show a consistent element association of Au-Ag-Cu-Pb-Zn ±Bi±W. The bulk Au/Ag ratios of the samples average 1.2 but they show a wide range due to the small size and variable mineralogy of the grab samples. The base metals Cu, Pb, and Zn occur at or above the 0.1% level in most samples. Sample 6134 from pyritized granite shows little

Table 3.1. Chemical composition of ore samples from the Niuxinshan deposit

Sample	6022	6023	6115	6119	6134	6153	6158	6159	6183	6186
Elements in wt %										
Cu	0.40	2.72	0.10	0.50	<0.01	0.03	0.84	0.01	1.04	0.05
Fe	4.1	6.6	4.6	4.3	21.6	3.8	13.7	12.2	20.0	11.8
Mn	0.3	0.1	0.1	0.1	0.1	0.1	0.1	0.1	0.1	0.2
Pb	0.34	0.39	0.04	1.29	0.11	0.10	0.66	3.94	3.89	0.62
S (total)	5.3	17.0	6.4	16.6	23.6	5.4	19.2	14.6	32.8	14.3
Zn	4.04	29.5	1.8	2.0	0.03	>2.0	>2.0	0.59	>2.0	>2.0
Elements in ppm										
Ag	25	62	16	86	324	36	51	11	166	41
As	7.0	6.0	25	1085	25	203	118	25	733	80
Au	6.6	50	21.8	5.6	732	4.7	63	13.7	66	179
Bi	27	56	44	203	380	114	96	8	292	55
Co	8	15	10	13	17	13	48	15	30	42
Cr	18	21	16	5	0.5	45	12	40	0.5	21
Mo	<2.5	<2.5	136	2.5	2.5	65	308	350	37	2.5
Ni	18	63	18	36	0.5	25	75	28	57	88
Sb	1.5	5.4	<5	<5	<5	<5	<5	<5	<5	<5
W	14	229	145	1700	621	1000	<1	143	200	146
Elements in ppb										
Pd	<1	<1	3	1	6	1	8	5	4	8
Pt	5	10	8	6	36	2	17	2	15	38

Analyses by Bondar-Clegg laboratories, Ottowa, Fire-assay/DCP (Au, Pd, Pt), all others by ICP.

enrichment in the base metals despite the extreme gold concentration of 730 g/t. The Bi concentration in this sample, like gold, is the highest of all the samples analyzed (380 ppm). The Au/Ag ratio is 2.2.

3.1.4 Wall Rock Alteration

Wall rock alteration is found around all mineralized veins in both the amphibolite and granite wall rocks. The igneous dikes may also produce intense wall-rock alteration in the amphibolites, and the effects of both dike and vein emplacement can be superimposed and difficult to distinguish. In the granite the total width of alteration zones reaches one or more meters and the altered rock locally contains ore grade mineralization. In the amphibolite, the visible alteration zone rarely reaches more than a few decimeters into the wall rocks and the altered rock is very weakly mineralized.

Alteration of Amphibolite
The wall rock alteration of amphibolite involves a bleaching of the rock due to the replacement of hornblende, biotite, and/or pyroxene by chlorite,

69

sericite, quartz, and carbonate. The bleached zones are commonly sheared and permeated by stringers of quartz and carbonate veinlets. Pyrite and lesser chalcopyrite form isolated grains in the mafic-rich portions of the wall rocks in association with hornblende and chlorite. In the vicinity of the Niuxinshan Granite and along some granitic dikes the altered amphibolite also contains red K-feldspar and fluorite. A late stage of carbonate occurs as cross-cutting veinlets which postdate other alteration minerals and the sulfide mineralization.

The chemical effects of alteration are shown in Fig. 3.5 which plots the changes in element concentration relative to the unaltered rock. The graphical "isocon" method of Grant (1986) was used to confirm that the relative gains and losses shown are not biased by a total change of mass. The alteration involved strong and consistent increases in the ore-forming elements Cu, Pb, Zn, W, Bi and, locally, Au. Elements leached from the rocks include Si, Na, the alkaline earths Ba and Sr (not shown), and also to some extent the rare earths and Y. The strong enrichments of S and CO_2 in the altered rocks show that the hydrothermal fluids were rich in these components. It may be the presence of carbonate in the fluid which caused the REE mobility (Taylor and Fryer 1984).

A special variety of wall rock alteration in the amphibolites occurs where quartz veins intersect layers of magnetite-rich amphibolite and magnetite quartzite. In these rocks the magnetite is replaced by pyrite. Advanced alteration in the magnetite-rich rocks leads to vein selvages up to 10 cm thick made up essentially of pyrite and quartz. These selvages locally contain gold values in the g/t range. The pyritization of magnetite in the wall rocks is a local but common phenomenon, and its possible significance for gold mineralization is discussed in detail in Chapter 4.3.

Alteration of Granite

The alteration in the Niuxinshan Granite is characterized by the replacement of the feldspars and biotite by quartz and white mica. Fluorite and pyrite are commonly present in the intensely altered samples. Pyrite may reach over 50 vol% of the rock, which then consititutes a rich gold ore with up to several hundred g/t Au. Carbonate is locally present, but it is later than the other secondary minerals.

The chemical effects of alteration (Fig. 3.6) involved the almost complete leaching of Na and local losses of Si and Ca. These were offset by gains of Fe, Mg, Mn, and K. Despite the destruction of plagioclase, the concentrations of Sr and Ba (not shown) increased during alteration. The altered granite is strongly enriched in S and in the chalcophile elements including gold. The "isocon" method of Grant (1986) was used to confirm that the relative changes shown are significant, since the alteration involved negligible changes in total mass of the system.

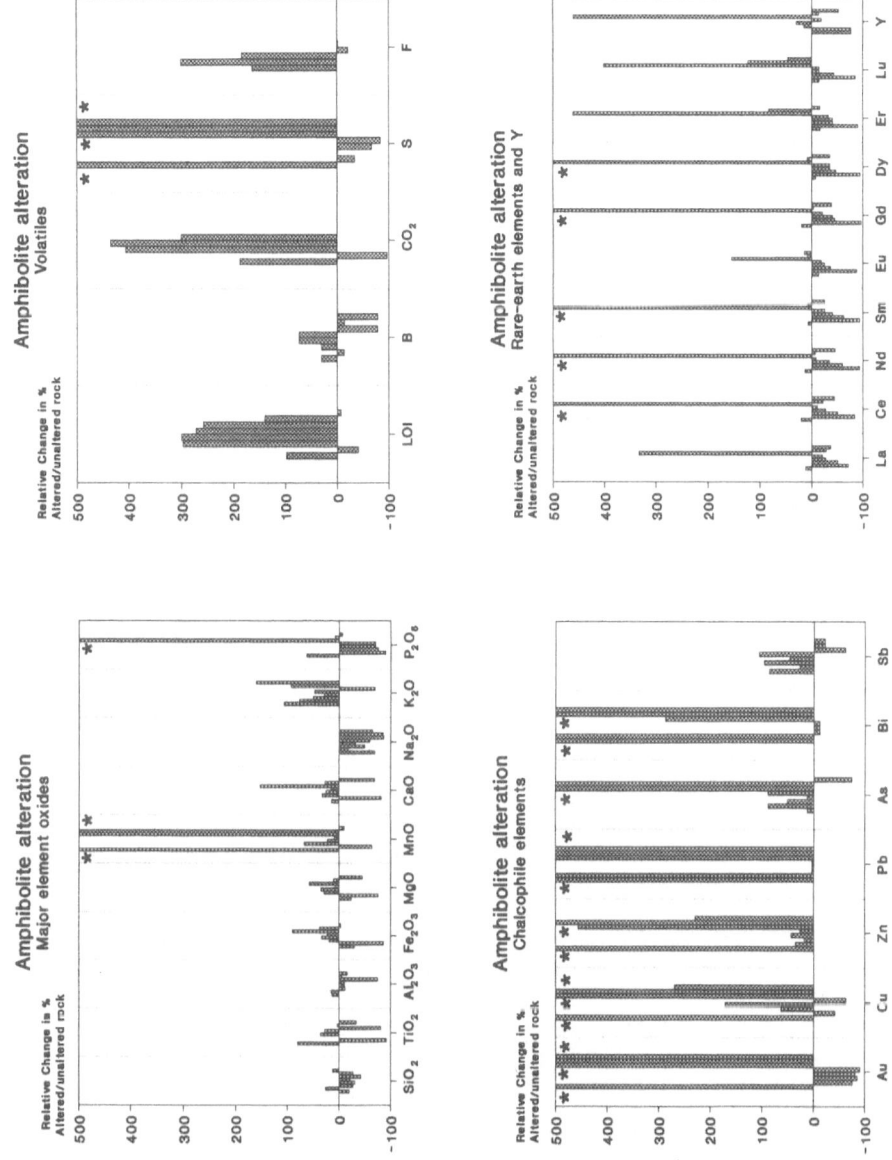

Fig. 3.5. Chemical changes accompanying wall rock alteration of amphibolites around ore veins from the Niuxinshan deposit. *Asterisks* denote element enrichments greater than 500%

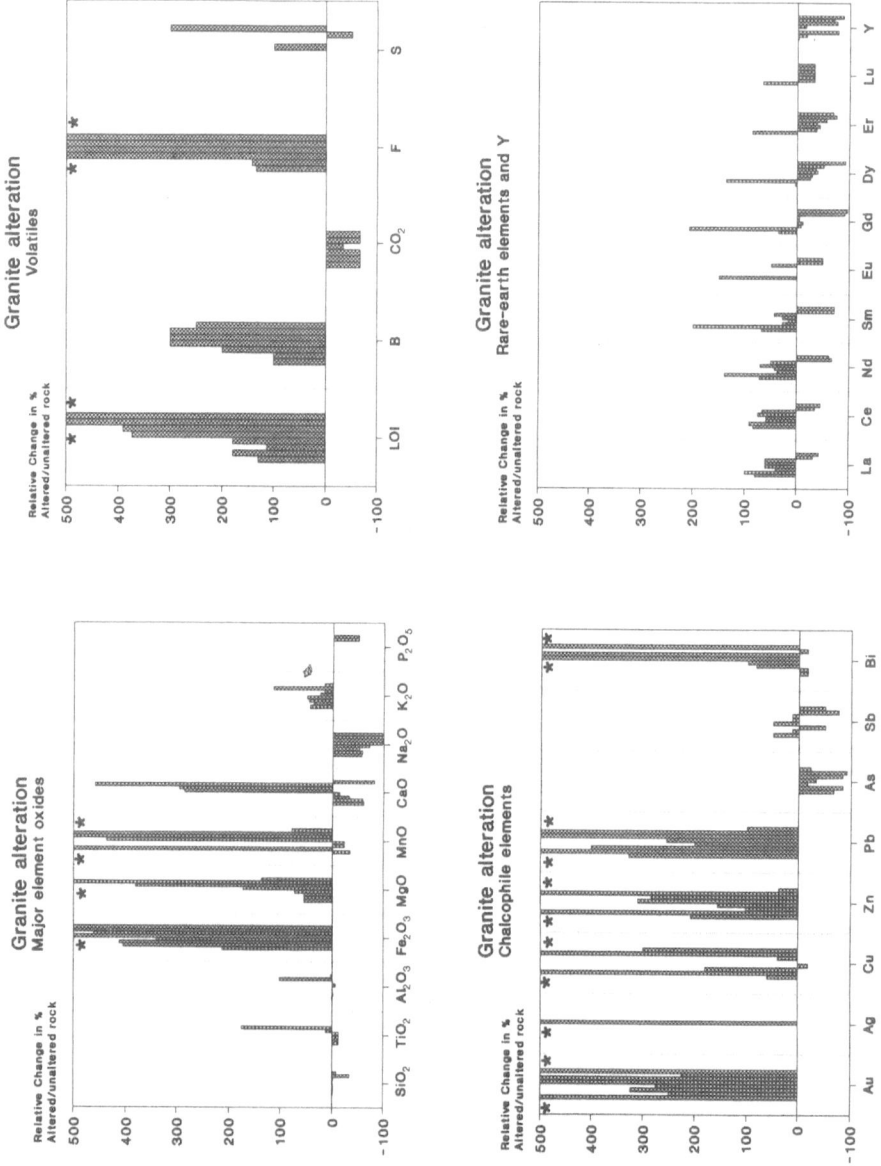

Fig. 3.6. Chemical changes accompanying wall rock alteration of the Niuxinshan Granite around ore veins, Niuxinshan deposit. *Asterisks* denote element enrichments greater than 500%

3.1.5 Age of Mineralization

The strongest field evidence constraining the age of mineralization is that the mineralized quartz veins cross-cut the Niuxinshan Granite and several of the post-granite igneous dikes, and are in turn cut by lamprophyre dikes. An absolute maximum age of the mineralization is therefore given by the age of the granite, which has been determined for the main phase (white granite) at 174 ± 14 Ma by the Rb-Sr isochron method. The dikes have not yet been extensively dated. Sun et al. (1989) reported K-Ar ages of 166 Ma from a lamprophyre from the Huajian deposit. The authors did not mention the relationship of the lamprophyre dike with ore veins, but the observations at the Niuxinshan deposit show that lamprophyres postdate the mineralization.

New age dating undertaken in this project from the Niuxinshan deposit involved Rb-Sr analysis of hydrothermal minerals and dating of vein quartz by the ^{40}Ar-^{39}Ar incremental heating method.

Rb-Sr Dating

Rb-Sr isotopic analyses were made from sericite and K-feldspar taken from the selvage zones of quartz veins. The analytical results are given in Table 3.2.

The best estimate for the age of mineralization based on these data is 175–190 Ma, from white mica (6038) and K-feldspar sample 6181. These ages are in good agreement with a K-Ar age of 188 Ma from hydrothermal sericite from the Huajian deposit in the same district reported by Yu and Jia (1989). They are also consistent, within the stated uncertainty, with the age of the Niuxinshan Granite. However, the sericite sample 6133 gives an unreasonably old age. The sample was taken from a mineralized zone in the Niuxinshan Granite, and yet it gives an apparent age nearly 100 Ma older than the host granite. The mica was apparently enriched in ^{87}Sr after its formation.

Table 3.2. Rb-Sr isotopic composition of hydrothermal vein minerals from the Niuxinshan deposit

Sample	Mineral	Rb	Sr	^{87}Rb/^{86}Sr	^{87}Sr/^{86}Sr[a]	Age[b]
6038	Sericite	620	5.56	352	1.6588(3)	190 ± 2
6133	Sericite	650	6.05	356	2.1780(4)	290 ± 2
6181	K-feldspar	508	283	5.21	0.7231(4)	175 ± 25

[a] Numbers in parentheses give uncertainty in the last digit (2 sigma).
[b] Ages in millions of years with uncertainty based on an initial ^{87}Sr/^{86}Sr ratio of 0.710 and a 1% relative error in ^{87}Rb/^{86}Sr.

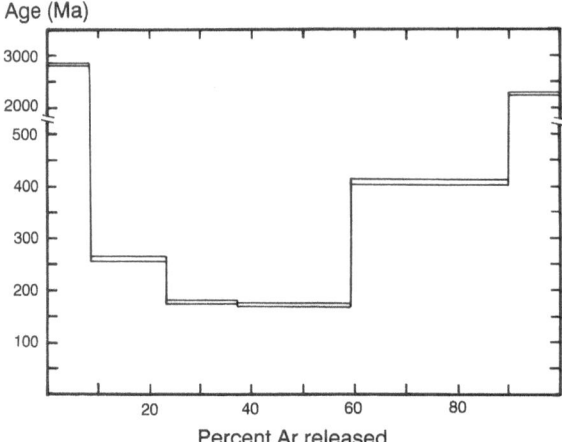

Fig. 3.7. Ar-Ar age-release curves of a sample of vein quartz from the Niuxinshan deposit. *Double lines* reflect analytical uncertainty. Analyses performed by the Ministry of Metallurgical Industry, Beijing

$^{40}Ar/^{39}Ar$ Dating

The results of $^{40}Ar/^{39}Ar$ dating of vein quartz from the Niuxinshan deposit by the incremental heating method are shown in Fig. 3.7. The analyses were performed by the Ministry of Metallurgical Industry, Beijing. The indicated plateau age of 176 ± 3 Ma agrees very well with the range of Rb-Sr ages given above and with the K-Ar age from the Huajian deposit.

3.1.6 Fluid Inclusions

Fluid inclusions from the Niuxinshan deposit were studied by both micro-thermometric methods and by chemical analyses of bulk fluid extracted by decrepitation. The nature of the samples was such that only fluorite and sphalerite were well suited for microthermometry, whereas the chemical analyses were performed on samples of vein quartz. This difference in the materials analyzed should be kept in mind when interpreting the results.

Microthermometry

Most samples of vein quartz from the Niuxinshan deposit are poorly suited for microthermometry because the quartz is obscured by clouds of very small secondary fluid inclusions. On the other hand, samples of fluorite and sphalerite from the ore veins contain large fluid inclusions well suited for microthermometry. Based on their large size, regular form, and random orientation in the host grains, the inclusions are interpreted to be primary

74

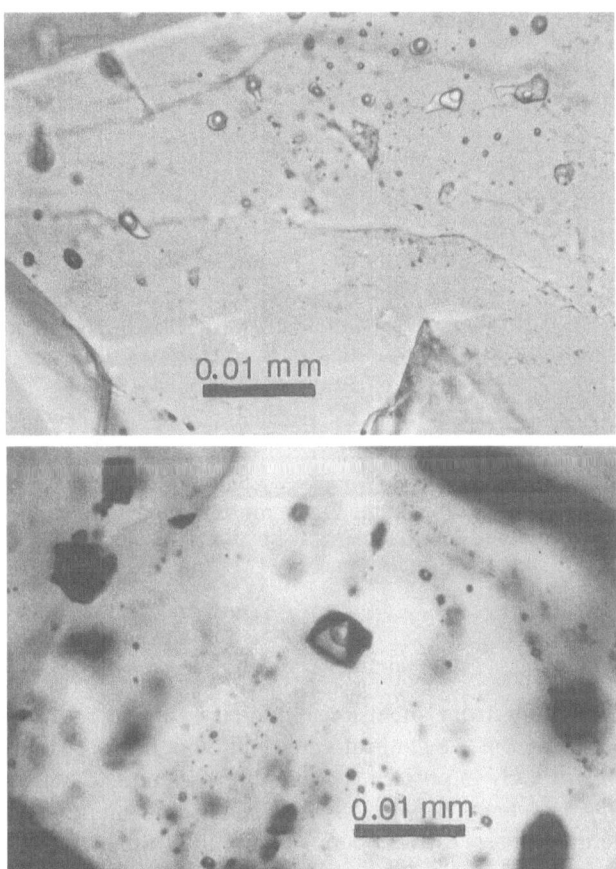

Fig. 3.8. Primary CO_2-H_2O fluid inclusions in fluorite (*above*) and in sphalerite (*below*) from the Niuxinshan deposit

(Roedder 1984). Secondary inclusions in these samples, recognized by their alignment along microcracks, were observed but not measured.

All samples measured come from quartz veins in amphibolite host rocks. The fluorite occurs in the altered vein selvages. In both fluorite and sphalerite the primary inclusions are rich in CO_2, and most inclusions show three phases at room temperature (H_2O liquid, CO_2 liquid, and gas). The degree of fill is about 70%. Examples are shown in Fig. 3.8. Secondary inclusions are generally smaller, have lower gas/liquid proportions, and show no visible liquid CO_2 at room temperature.

Table 3.3 gives the microthermometric data from three samples of fluorite and two of sphalerite. In general, the results from the two host minerals are very similar. The melting temperatures of CO_2 below $-56.6\,°C$ indicate some impurity of this phase, as is also suggested by the chemical analyses of

Table 3.3. Summary of microthermometric data from fluid inclusions from the Niuxinshan deposit

Host mineral	Sample	T_m CO_2	T_m clathrate	T_h CO_2	T_h final
Fluorite	6019-1	n.d.	7.9	n.d.	273
	(n = 20)	–	(0.1)	–	(3.5)
	6019-1b	−57.2	8.2	29.3	313
	(n = 34)	(x)	(0.4)	(0.2)	(5.2)
	6019-1a	n.d.	7.9	n.d.	315
	(n = 29)	–	(0.2)	–	(3.9)
Sphalerite	6018a1	−59.0	6.4	21.3	266
	(n = 6)	(x)	(0.2)	(1.9)	(6.4)
	6018a2	−58.3	6.6	27.6	>280
	(n = 5)	(0.1)	(0.1)	(1.0)	–

All temperatures represent average values in °C.
T_m, melting temperature; T_h, homogenization temperature; n.d., not determined; (x), number of measurements to small for statistical analysis.
The number of measurements (n) and the standard deviation are given in parentheses.

bulk fluid from inclusions in quartz discussed below. The estimation of salinity from the clathrate melting temperatures is only approximate because of the CO_2 impurity. A minimum estimate, based on the pure CO_2-H_2O-NaCl system, is 3 to 7.5 wt% NaCl equivalent. The final homogenization temperatures for inclusions in fluorite range from 275 to 315 °C (homogenization to the liquid phase or critical). The homogenization temperature of inclusions in sphalerite could rarely be measured because of decrepitation. Some inclusions homogenized to the liquid phase at 266 °C ± 6 °C but most inclusions decrepitated before homogenization at temperatures of 270 to 280 °C. Based on the similarity of inclusion types between sphalerite and fluorite, it is likely that the homogenization temperature in sphalerite would also be around 300 °C.

These temperatures represent a minimum estimate of trapping conditions because no pressure correction was applied. From the position of the CO_2-H_2O solvus at the appropriate salinity given by Brown and Lamb (1989) a minimum trapping pressure of about 1.5–2 kbar can be derived from these data. The true conditions of trapping may be close to 430–490 °C and 3.5–5 kbar, based on oxygen isotopic thermometry of vein quartz and K-feldspar (see below) combined with the fluid isochores calculated assuming pure CO_2-H_2O-NaCl from Brown (1989).

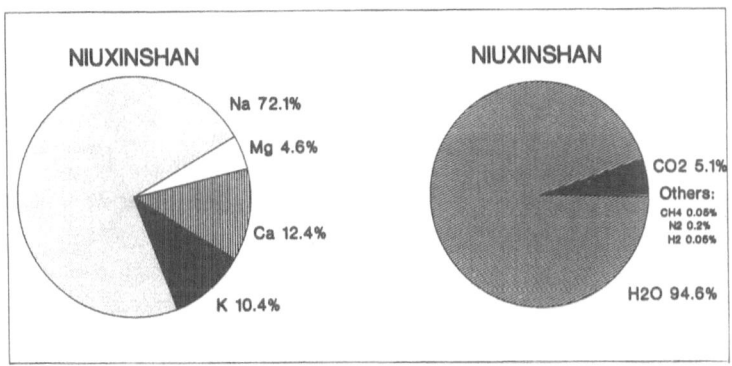

Fig. 3.9. Pie diagrams showing the average composition of bulk fluid inclusion contents from vein quartz in the Niuxinshan deposit. From unpublished analyses of the Ministry of Metallurgical Industry, Beijing

Chemical Analyses

Chemical analyses were performed by laboratories of the Ministry of Metallurgical Industry in Beijing. Bulk fluid was released from the inclusions by heating to decrepitation in vacuum. Dilution from secondary inclusions was minimized by analyzing only the fluid fraction released above 250 °C, after decrepitation of most secondary inclusions. Nevertheless, the results are sure to reflect some mixture of different fluid generations and they cannot be directly compared with the microthermometric data, which was obtained from selected primary inclusions in different host minerals.

Figure 3.9 shows the average fluid composition in terms of the proportion of dissolved cations and gas species. The complete data are contained in an internal project report to the European Community and may be obtained from the authors on request or from the European Community (1990). Sodium makes up about 72% of the dissolved cations. Potassium and Ca each constitute an additional 10%, and Mg is the least abundant. The most abundant gas species is CO_2, with about 5 mol%. Other detectable gases together make up less than 0.5 mol%.

3.1.7 Stable Isotopic Data

This section presents new data on oxygen and carbon isotopic composition of gangue minerals from ore veins in the Niuxinshan deposit. The main purpose of the analyses was to provide estimates of the temperature of mineralization, and only this aspect is discussed here. The implications of the isotopic data to the source of mineralizing fluids is discussed in Chapter 4.6.2, together with data from the other deposits.

Table 3.4. Oxygen and carbon isotopic composition of quartz, K-feldspar, and carbonate from hydrothermal veins in the Niuxinshan deposit

Sample	Mineral	δ^{13}C PDP	δ^{18}O SMOW	Temperature (°C) Ksp-Qtz[a]
6020	K-feldspar	–	9.2	–
6020	Quartz	–	10.9	470
6111	K-feldspar	–	8.7	–
6111	Quartz	–	10.6	430
6181	K-feldspar	–	10.0	–
6181	Quartz	–	11.6	493
6555	K-feldspar	–	11.2	–
6555	Quartz	–	11.7	1098
6173	Dolomite	−6.3	9.7	–

[a] Temperatures calculated from quartz-alkali feldspar fractionation factors of Clayton et al. (1989).
The analytical precision of ±0.2‰, corresponds to a temperature uncertainty of ±70 °C.

Table 3.4 lists the isotopic data and the temperatures calculated from quartz-K-feldspar pairs. Four samples of vein quartz from the Niuxinshan deposit show a narrow compositional range of δ^{18}O (SMOW) from 10.6 to 11.6‰. The K-feldspar samples have more variable isotopic composition, from 8.7 to 10.2‰. The quartz is consistently enriched in ^{18}O relative to coexisting K-feldspar, which is consistent with isotopic equilibrium (O'Neil 1986). The temperatures calculated from the quartz-K-feldspar fractionation factors (based on Clayton et al. 1989) from three of the four mineral pairs fall in the range of 430 to 490 °C (note the high uncertainty of ±70 °C), and this is further evidence for isotopic equilibrium. The fourth mineral pair yields an impossibly high temperature (1098 °C).

These temperatures are 100 to 150 °C higher than the fluid inclusion homogenization temperatures. This difference may be reasonable since the true trapping temperatures of the fluid inclusions are expected to be higher than their homogenization temperatures, which represent only minimum estimates. On the other hand, the P-T conditions suggested by isotopic thermometry combined with fluid isochores (430–490 °C, 3.5–5.5 kbar) indicate upper greenschist facies conditions, and this is inconsistent with the brittle deformation style of the veins.

3.2 Sanjia District

The Sanjia gold mining district is located about 250 km northeast of Beijing in Qinglong county, Hebei province (see Fig. 3.1). There are presently (as of 1989) three producing gold deposits in the district, namely, Sanjia,

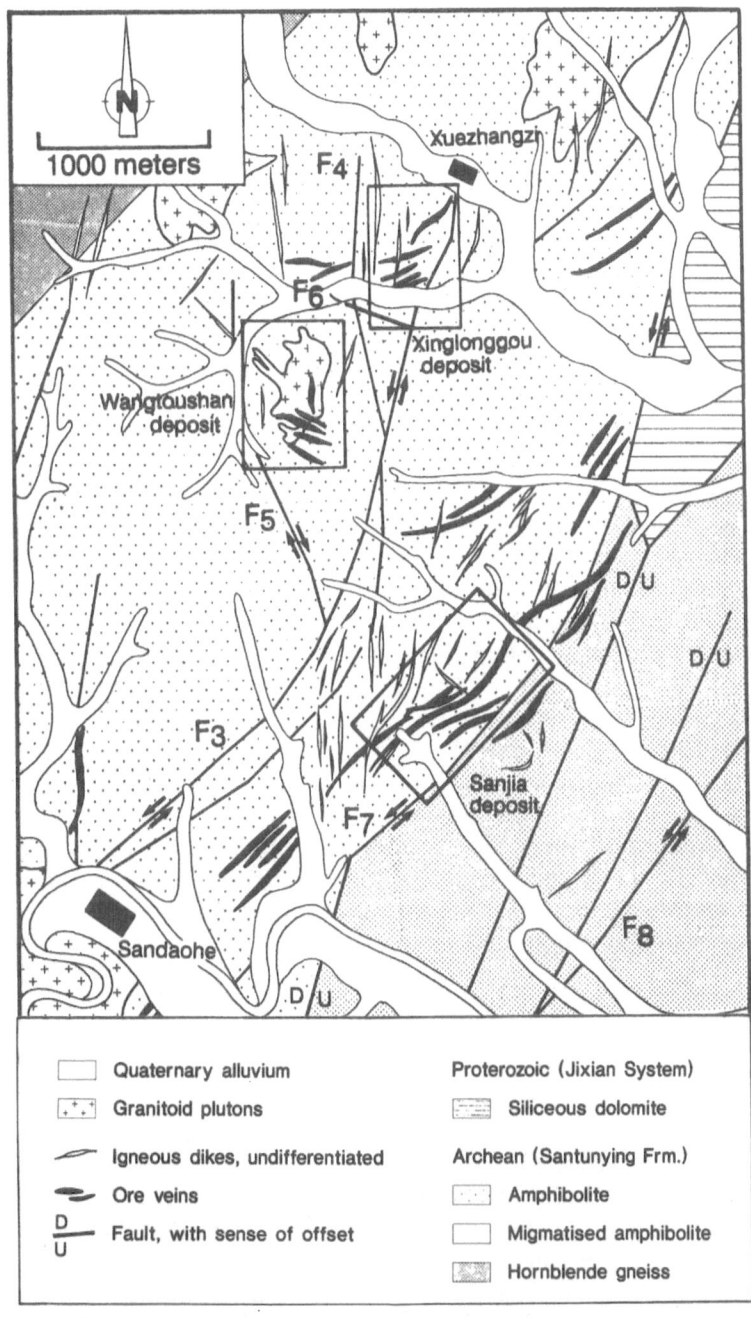

Fig. 3.10. Simplified geologic map of the Sanjia district. The locations of the Sanjia, Wangtoushan, and Xinglonggou gold deposits are *boxed*

Wangtoushan, and Xinglonggou, all of which are described in some detail in this text. The location of these deposits is shown in a geologic map of the district in Fig. 3.10.

The Sanjia Deposit
The Sanjia deposit was discovered during the Qing dynasty (1644–1911 A.D.). The gold ore occurs in quartz veins within amphibolite host rocks of the Archean Qianxi Group. The deposit contains 20 ore-bearing veins, the largest and most productive of which is about 1 km long and known to a depth of 500 m. The veins range in thickness from a few centimeters to 1 m. Ore bodies have gold concentrations of about 20 g/t on average. The Sanjia deposit is mined by the Qinglong county government. The workings are entirely underground. The ore treatment is done in a central dressing plant at Sanjia village.

The Wangtoushan Deposit
The Wangtoushan deposit was discovered in 1976. Gold ore occurs in quartz veins within Archean amphibolites of the Qianxi Group and, to a lesser extent, within a granite intrusion. Six gold-quartz veins are worked in an underground operation. The ore grade averages about 30 g/t. The mine is run by Qinglong county. Ore is treated at the central dressing plant in Sanjia.

The Xinglonggou Deposit
This gold deposit was first discovered in 1969. Like the other deposits in the district, the ore occurs in quartz veins within Archean amphibolites. About ten veins are known. The average ore grade is 23 g/t Au. The Xinglonggou mine is run by the Qinglong County government and is one of the most promising of the newer gold prospects in the Sanjia district. The workings are entirely underground.

3.2.1 Host Rock Lithology

The main host rocks in the Sanjia district are Archean amphibolites and gneisses of the Santunying Formation, Qianxi Group. In the northeastern part of the district Middle Proterozoic sedimentary rocks (Jixian and Changcheng Systems) are in fault contact with the Archean rocks (Fig. 3.10), but the Proterozoic rocks do not constitute a host to ore. Igneous rocks are represented by several granitic intrusions of presumed Yanshanian age, and by numerous felsic and mafic dikes, most of which are younger than the granites and older than the mineralization.

3.2.1.1 Archean Rocks

The Archean Santunying Formation consists of a complex series of amphibolites which are variably migmatized. In the geologic map of the Sanjia district (Fig. 3.10) the Santunying Formation is divided into a lower, middle, and upper section. Note that this geologic division is local and it may not be valid regionally.

The structurally lowest unit (fine stippled pattern in Fig. 3.10) consists of highly migmatized interlayered hornblende amphibolites and biotite amphibolites. The migmatization is expressed by schlieren and veinlets of leucosome and/or K-feldspar blastesis. Migmatization is strongest in the lower parts of the unit and decreases upwards. The middle unit (coarse stipple in Fig. 3.10) consists of nonmigmatitic layered hornblende amphibolites and biotite amphibolites. Most of the mines in the Sanjia district are located in this section of the Santunying Formation. The upper section of the Santunying Formation consists of a variety of amphibolites and hornblende-plagioclase gneisses which occur at the northwest corner of the area (shaded pattern in Fig. 3.10) and do not constitute a host to ore.

Important minor rock types within the Santunying Formation include quartz-magnetite amphibolites which grade into magnetite quartzites, and hornblendites. The layers of quartz-magnetite amphibolite range in thickness from a few centimeters to about 1 m and are strictly parallel to the surrounding rock structure. They occur in all sections of the Santunying Formation but seem most abundant in the rocks of the middle section in the area of the Sanjia deposit. The hornblendites occur as discontinuous lenses, some showing boudinage structure, or as isolated xenoliths in the amphibolites of all sections of the Santunying Formation. These ultramafic rocks probably represent relics of former dikes in the protolith sequence.

All of the rocks in the Santunying Formation are strongly foliated, with foliation parallel to lithologic layering, and small-scale isoclinal folds are commonly seen in outcrop. Details of the structure are discussed in Chapter 3.2.2.

3.2.1.2 Granite and Dikes

Several intrusive stocks of presumed Yanshanian age intrude the Archean basement in the Sanjia district (Fig. 3.10). Two of these intrusions (Wangtoushan and Sanyihe) are locally mineralized although the gold concentrations are subeconomic. The Wangtoushan Granite is directly associated with the Wangtoushan gold deposit. The Sanyihe Granite is located about 3 km southwest of the Sanjia deposit.

The Wangtoushan Granite is a small pluton of medium-grained leucocratic pinkish granite (outcrop area 0.25 km^2) with several NE–SW- and NW–SE-striking apophyses. On the southeastern border of the pluton is a breccia zone consisting of decimeter- to meter-size angular blocks of granite cemented by quartz and graphic granite/pegmatite. The granite consists of approximately 40% albite, 40% K-feldspar, 20% quartz, and 1–2% biotite. The chemical composition reflects the highly leucocratic nature of the granite, with an average 76 wt% SiO_2, 9 wt% total alkalis (K_2O/Na_2O near 1), and the sum of total Fe as FeO, MgO, MnO, CaO, and TiO_2 is less than 1.5 wt%. The concentrations of Ba (average 185 ppm), Sr (average 13 ppm), and Rb (average 247 ppm) are typical for a moderately differentiated granite. The chondrite-normalized REE pattern (not shown) has a large negative Eu anomaly and nearly equal light and heavy REE enrichments. A summary of analytical data is given in Appendix 1.

The Sanyihe Granite forms a small intrusion (about 3 km^2 in outcrop) on the southwestern edge of the Sanjia district. The light gray granite is medium-grained, equigranular, and consists of approximately 35% oligoclase, 40% K-feldspar, 20% quartz, and 5% biotite, with accessory titanite, rutile, apatite, and primary white mica. The chemical composition is very similar to that of the Wangtoushan Granite summarized above, although it is slightly less leucocratic and has correspondingly higher Fe, Mg, and Ti concentrations (see Appendix 1).

Igneous dikes of both mafic and felsic composition are abundant in the Sanjia district. Three groups of dikes are distinguished: porphyritic diorite, granitic (with aplitic or pegmatitic texture), and lamprophyre. All three groups intrude along steeply westward-dipping NE–SW- and N–S-trending faults and fracture zones. The most prominent dikes in terms of size are the granitic dikes, which are typically several meters thick and are exposed over several hundred meters along strike. In the Wangtoushan area it is clear that many of the granitic dikes are apophyses of the Wangtoushan Granite, and it is possible that the other dikes in the district are likewise connected to larger granitic intrusions at depth. Field relations show that all of the dikes cut the granitic stocks at the present erosional level. Cross-cutting relations among the different dikes are not abundant. In some cases, the dioritic dikes are cut by granitic dikes although relations to the contrary can also be found. Both the dioritic and granitic dikes are cut by the ore veins. It is clear that the lamprophyre dikes are the latest intrusions, and they also clearly postdate the mineralization. The largest dikes are shown in Fig. 3.10 but are not differentiated by composition.

3.2.2 Host Rock Structures

The Archean basement rocks in the Sanjia district are isoclinally folded, and the associated axial-planar foliation strikes NE–SW and dips steeply NW. Extensive systems of fracture and fault zones postdate the folding. The granites and dikes intruded after the folding episode and are affected only by brittle deformation.

3.2.2.1 Folds and Foliation

The Sanjia district is located on the eastern limb of the NNE trending Dushan Anticlinorium which, according to Sun et al. (1989), belongs to the early generation of folds, i.e., Archean/Early Proterozoic (see Chap. 2.3.1). The folds of this generation dominate the structure of the Archean basement rocks in the district. They are tight to isoclinal folds overturned to the east with northwest-dipping axial planes. The foliation in the amphibolites and gneisses is axial planar to these folds, strikes generally NE–SW and dips 50°–70° to the NW. The foliation is parallel with lithologic layering and it postdates the migmatization, i.e., the leucosomes are also folded and foliated with the surrounding rock.

3.2.2.2 Faults and Fractures

There are two main orientations of faults and fracture zones in the Sanjia district, namely NE–SW and N–S. Most of the mineralization and most of the igneous dikes are associated with the NE- to NNE-trending structures. A few faults of NW–SE trend also occur, and many of these are later than the previously mentioned structures and also postdate the mineralization.

NE–SW-Trending Faults
A number of NE–SW-trending faults extend through the entire district and these are designated on the geologic map (Fig. 3.10) as F3, F7, and F8 (no age sequence is implied by the numbers). This group of faults strikes 20°–35° and dips 50°–80° to the NW. They are sinistral oblique-slip faults which consist of brecciated and sheared zones several meters to tens of meters wide with gently undulating fault surfaces. Most of the ore veins in the Sanjia and Xinglonggou deposits occur in faults interpreted to be secondary splays of the F3 and F7 faults.

N–S-Trending Faults
Faults and fractures with a N–S strike direction have strike-slip characteristics or are tensional in nature. Many of the tensional N–S fault zones have been intruded by dikes (Fig. 3.10). The main map-scale fault with a N–S orientation is the F4 fault which dips 70°–80° to the west and is characterized by a brecciated zone which is locally silicified and mineralized.

NW–SE-Trending Faults
NW–SE-trending faults and fractures have been divided into two groups according to their relative age. The earlier group of faults are dextral strike-slip faults interpreted as conjugate faults to the sinistral NE–SW-trending faults discussed above. A map-scale example is the F5 fault (Fig. 3.10). The later group of NW-trending faults cuts igneous dikes and ore-bearing veins.

An example is the F6 fault, which is a strike-slip fault striking 300° and dipping 65° to the SW.

3.2.3 Gold Mineralization

The gold mineralization in the Sanjia district occurs in quartz-sulfide veins which cross-cut both amphibolite and also Yanshanian Granite (Wangtoushan and Sanyihe Granites); however, only those veins within amphibolite wall rocks contain economic ore bodies. The style of mineralization and the ore paragenesis in the Sanjia, Wangtoushan, and Xinglonggou deposits are very similar. The following descriptions are generalized from observations of all three deposits except where stated otherwise.

3.2.3.1 Form of the Ore Bodies

All of the economic ore bodies in the Sanjia district occur in veins within Archean metamorphic rocks. Where the mineralized quartz veins cut the Wangtoushan and Sanyihe Granites, they splay out into thin veinlets in a zone of pervasively altered (sericite-quartz-pyrite) granite. These zones contain only subeconomic gold values.

In the *Sanjia deposit* the ore-bearing veins strike 60–70° and dip 25–50° to the NW. The host fault surfaces are planar and display indications of dextral movement. These faults are considered to be secondary splays of the main F7 fault (see Fig. 3.10) although the sense of offset on F7 is sinistral.

In the *Xinglonggou deposit* the ore-bearing veins also occupy fault zones showing evidence of shear. Unlike Sanjia, both NE–SW faults with dextral shear and NW–SE faults with sinistral shear occur, and these are interpreted as a conjugate pair of structures related to E–W-directed stress.

In the *Wangtoushan deposit* most of the veins occur in WNW-trending tensional faults which show only minor shearing. The ore-bearing veins have a maximum length of 600 m and a width of 10 to 60 cm. The veins dip 10–45° NE. One NNW-trending gold-bearing quartz vein also contains an economic ore body.

3.2.3.2 Macroscopic Description of the Ores

Because only local, subeconomic mineralization occurs in the granites of the Sanjia district, the following discussion is confined to ores in amphibolite wall rocks. The mineralization typically forms in milky white massive vein quartz. The veins pinch and swell and show evidence of multiple re-opening. Rich parts of the veins may be strongly brecciated and cemented with pyrite and base metal sulfides. The wall rocks are generally highly sheared and

Fig. 3.11. *Above*: Photograph of ore veins in underground workings of the Wangtoushan deposit. *Below*: Photomicrograph of ore from the Xinglonggou deposit. Tetrahedrite (*te*), chalcopyrite (*cp*), and galena (*ga*) occur with gangue (*black*) in fractured pyrite (*py*)

altered but rarely are they significantly mineralized except directly at the vein contact. Slivers of wall rock are common in the veins near the walls. The sulfide minerals occur along the vein selvage and in seams and nests within the veins. Locally, an early mineralization by pyrite can be distinguished macroscopically from a later pyrite-galena-sphalerite mineralization based on different generations of vein opening and filling. The typical sulfide concentrations in the mineralized portions of the veins is 5 to 10 vol%. Carbonates typically occur as late-stage gangue minerals filling cross-cutting fractures or disseminated in the altered wall rocks. A typical example of ore veins in amphibolite host rock is shown in Fig. 3.11.

The veins at the Wangtoushan deposit have some special features which are not found in the Sanjia or the Xinglonggou deposits. Reddened K-feldspar commonly occurs in the quartz veins near the wall rock selvages, and in some cases fluorite accompanies the K-feldspar in the veins. More often, fluorite occurs in the altered selvage zones. A third mineralogical peculiarity of the Wangtoushan deposit is the occurrence of accessory molybdenite. These features are thought to reflect the proximity of the Wangtoushan Granite.

3.2.3.3 Ore Petrography and Paragenesis

The ore paragenesis in the three deposits of Sanjia, Xinglonggou, and Wangtoushan is nearly identical. All three deposits contain chalcopyrite, galena, native gold, pyrite, sphalerite, and tetrahedrite. Pyrrhotite and native bismuth were found only at the Sanjia deposit, covellite occurs as a secondary mineral at Xinglonggou, and the mineralization at Wangtoushan contains accessory molybdenite and scheelite.

The following descriptions apply in general to the Sanjia district except where otherwise indicated.

Pyrite
Pyrite is the main sulfide mineral and it occurs in both the quartz veins and altered wall rocks. In the wall rocks pyrite occurs as scattered grains associated with mafic minerals (chlorite, hornblende), and it contains numerous silicate inclusions. In the veins pyrite is the oldest sulfide mineral and it occurs in grains and grain aggregates with individual crystals up to 5 mm in size. Locally, pyrite forms nearly monomineralic seams of some centimeters thickness in vein quartz. The pyrite is strongly fractured and filled by galena, chalcopyrite, tetrahedrite, and quartz (Fig. 3.11). Pyrrhotite (Sanjia deposit only) and chalcopyrite occur rarely as inclusions of a few microns in diameter. A younger pyrite generation occurs together with carbonate in late-stage veinlets which cut the earlier ore and gangue minerals. This late pyrite generation is not associated with other sulfides.

Chalcopyrite, Sphalerite, and Galena
Chalcopyrite is found almost exclusively in the veins with only rare grains dispersed in the wall rock selvages. Chalcopyrite in the veins commonly occurs intergrown with galena, sphalerite and tetrahedrite in fractured pyrite. Chalcopyrite is also common as exsolution blebs and inclusions in sphalerite. Rarely, fracture fillings of massive chalcopyrite with inclusions of galena occur in quartz. Finally, a late stage of chalcopyrite may be found locally in fractures with carbonate. At the Xinglonggou deposit, covellite locally rims chalcopyrite in weathered samples.

Sphalerite is not abundant in the Sanjia district. The mineral most commonly forms as fracture fillings in pyrite together with galena, chalcopyrite,

86

and tetrahedrite. Isolated grains and grain aggregates of sphalerite occur in quartz, and sphalerite exsolution starlets in chalcopyrite also occur.

Galena occurs mainly in fracture fillings in pyrite and quartz. Galena veins sphalerite and is intergrown with tetrahedrite and chalcopyrite. Native gold is frequently associated or intergrown with galena.

Native Gold

Native gold occurs mainly intergrown with galena, chalcopyrite, and tetrahedrite in fractures within pyrite. Minor amounts of gold are also found in fractures within quartz. Gold occurs locally as inclusions in pyrite, tetrahedrite, and even sphalerite. The size of the gold grains is mostly between 0.1 and 0.3 mm with rare coarse aggregates of up to 5 mm in size. SEM-EDS analysis of gold grains showed three compositional groups as follows:

- gold grains associated with sulfides in fractured pyrite have 35–45 wt% Ag,
- isolated gold inclusions in pyrite have concentrations of 20–30 wt% Ag,
- gold in cracks of quartz show the highest Ag concentrations of approximately 55 wt%.

Tetrahedrite

Tetrahedrite typically occurs intergrown with galena, chalcopyrite, and sphalerite in cracks of pyrite and quartz. It overgrows and veins chalcopyrite and galena and is therefore clearly younger than those minerals. The size of the tetrahedrite aggregates can reach several millimeters.

Pyrrhotite

Pyrrhotite is a rare mineral and it was found only in minor amounts as some micron-sized inclusions in pyrite at the Sanjia deposit.

Molybdenite

Molybdenite occurs only in the Wangtoushan deposit, where it forms irregular, felty aggregates of up to 2 mm in diameter in the quartz veins within and near the Wangtoushan Granite, and in the altered wall rocks associated with fluorite, carbonate, and rutile. In rare cases, molybdenite occurs in cracks of pyrite.

Native Bismuth

Native bismuth was found in samples from the Sanjia deposit as micron-sized inclusions in galena.

3.2.3.4 Chemical Composition

The compositions of 17 grab samples of well-mineralized ore taken from the Sanjia district are compiled in Table 3.5. The purpose of the analyses is to

Table 3.5. Chemical composition of ore samples from the Sanjia district, eastern Hebei province

Deposit	Xinglonggou				Wangtoushan									Sanjia			
Sample	6002	6003	6293	6294	6088	6262	6264	6272	6305	6314	6212	6214	6222	6223	6225	6227	6246
Elements in wt %																	
Cu	0.10	0.10	0.52	0.64	0.08	0.01	<0.01	0.01	2.56	0.72	0.59	0.18	0.24	2.35	0.99	0.59	0.08
Fe	27.1	26.7	10.8	9.6	29.5	18.7	25.8	15.8	5.4	30.4	4.6	3.4	8.2	11.8	7.1	2.6	15.1
Mn	<0.01	<0.01	0.01	0.05	<0.01	0.04	0.03	0.08	0.29	0.03	0.49	0.36	0.18	0.21	0.75	0.26	0.05
Pb	0.56	0.42	0.84	1.27	0.37	0.68	0.34	0.29	6.88	0.67	0.47	1.02	0.61	18.3	2.91	6.96	0.35
S (total)	22.0	24.0	11.7	11.5	25.0	19.3	30.1	17.9	12.0	34.7	10.7	3.5	10.1	22.4	10.8	8.3	17.1
Zn	0.01	0.01	0.02	0.02	0.07	0.25	<0.01	<0.01	12.68	0.18	<0.01	<0.01	12.8	2.0	<0.01	2.0	<0.01
Elements in ppm																	
Ag	133	133	158	488	140	298	200	228	98	60	31	51	51	139	126	55	186
As	13	50	55	607	10	25	25	25	311	25	453	25	25	353	353	269	25
Au	19	27	31	81	27	0.7	66	13	25	30	31	23	48	99	54	39	173
Bi	170	160	179	404	320	525	448	484	126	94	24	72	95	178	240	57	760
Co	16	20	22	17	35	14	12	19	30	39	14	18	16	54	42	13	22
Cr	21	23	3	1	6	0.5	0.5	15	11	0.5	40	60	7	13	4	2	4
Mo	<5	<5	18	14	110	704	<5	239	<5	14	2.5	2.5	21	2.5	2.5	2.5	2.5
Ni	10	14	30	50	33	19	24	24	91	52	45	117	36	300	73	32	40
Sb	5.6	16	<5	23	2.6	<5	<5	<5	<5	<5	<5	<5	<5	123	<5	40	<5
W	1.5	1.5	15	34	1.5	31	12	13	100	30	100	14	23	100	100	100	24
Zn	–	–	–	–	–	–	25	36	–	–	16	29	–	–	12	–	33
Elements in ppb																	
Pd	<1	<1	4	5	<1	2	5	6	3	3	3	3	4	6	4	4	8
Pt	5	5	14	25	10	2	23	32	13	16	15	12	15	30	17	15	37

Analyses by Bondar-Clegg laboratories, Ottowa. Fire-assay/DCP (Au, Pd, Pt), all other elements by ICP.

reveal typical element associations in the ore and to give a rough idea of the concentrations; they are not necessarily representative for the bulk deposits. Characteristic element associations of the ores in the Sanjia district are Au-Ag-Cu-Pb-Bi ±Zn ± Sb ±W. Mo is significant only in the Wangtoushan deposit, where molybdenite occurs in the veins. Gold concentrations in the Sanjia district reach over 100 g/t, but values of 20 to 30 g/t can be considered typical. The Ag/Au weight ratios are invariably greater than one, and with a few exceptions the ratios range from 1 to 2 in the Sanjia deposit, up to 4 to 8 in the other deposits. The base metals Cu and Pb are each present in concentrations of greater than 0.1 wt%. The Zn concentration is usually less than 200 ppm, although it reaches several percent in exceptional samples.

3.2.4 Wall Rock Alteration

Wall rock alteration is well developed around the veins in both the amphibolite and granite wall rocks. The igneous dikes also produce intense wall rock alteration, and the effects of both dike and vein emplacement may be superimposed.

Alteration of Amphibolite
The alteration of amphibolite is characterised macroscopically by a bleaching of the rock due to the replacement of hornblende and/or biotite by chlorite, sericite, epidote, quartz and carbonate. The bleached zones are commonly sheared and permeated by stringers of quartz and carbonate. Pyrite and rarely chalcopyrite form isolated grains in the mafic-rich portions of the wall rcks. In the Wangtoushan deposit the alteration zones also include red K-feldspar and fluorite.

The chemical changes brought about by alteration (Fig. 3.12) include strong and consistent increases in the ore-forming elements Cu, Pb, Zn, W, Bi, Au, and in the volatiles S and CO_2. Elements leached from the rocks include Si, Na, Fe, Mg, and also the rare earths and Y. The graphical "isocon" method of Grant (1986) was used to confirm that the alteration involved negligible changes in total mass of the system, and therefore the relative changes shown are significant.

A special variety of wall rock alteration in the amphibolites affects layers of magnetite-rich amphibolite and magnetite quartzite. In these rocks the magnetite is replaced by pyrite and in strongly pyritized rocks the gold values may reach into the g/t range. The pyritization of magnetite in the wall rocks and its significance for gold mineralization is discussed in detail in Chapter 4.3.

Alteration of Granite
The Wangtoushan and Sanyihe Granites show the same type of alteration where they are cut by mineralized quartz veins. The alteration is charac-

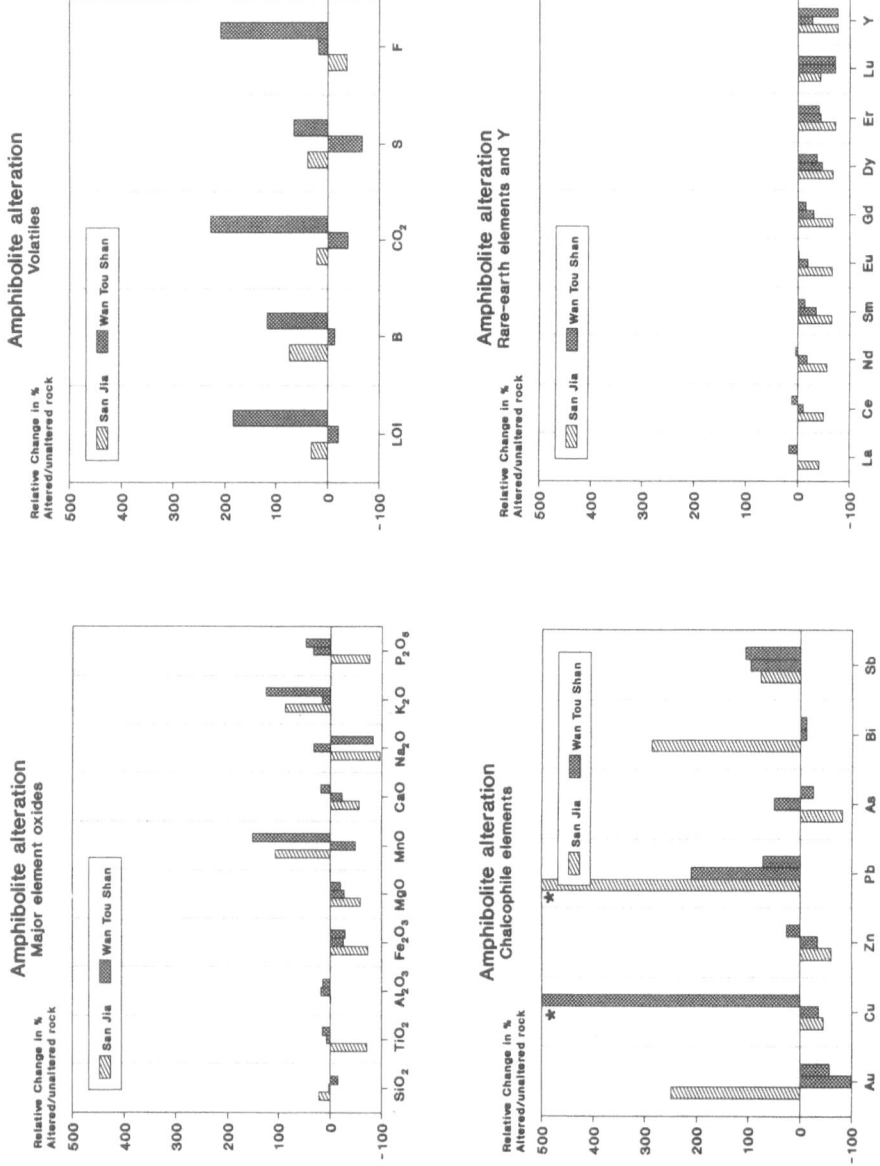

Fig. 3.12. Chemical changes accompanying wall rock alteration of amphibolite around ore veins from the Sanjia and Wangtoushan deposits. *Asterisks* denote element enrichments greater than 500%

terized by silicification and sericitization of both plagioclase and K-feldspar with minor growth of disseminated submillimeter sized euhedra of pyrite. Pyrite rarely reaches more than accessory concentration. Fluorite is rarely developed in the altered granite. The alteration extends outward up to 1 or 2 m from a swarm of quartz veinlets. The maximum gold concentration, according to exploration work done on these zones, is 4 g/t. The chemical effects of alteration (not shown on Fig. 3.12) include a loss of Na, Ca, Ba, and Sr; and the gain of Fe, Mg, S, F, W and the chalcophile ore elements relative to the unaltered granite.

3.2.5 Age of Mineralization

The field evidence in the Sanjia district indicates that the mineralized veins formed after the emplacement of the Wangtoushan and Sanyihe Granites and their associated dikes, and before the intrusion of lamprophyre dikes. The only isotopic age data available from the granites is an unpublished Rb-Sr whole rock isochron of the Sanyihe Granite obtained by the Ministry of Metallurgical Industry (European Community 1990). The isochron age of the Sanyihe Granite is 166 ± 2 Ma and the initial $^{87}Sr/^{86}Sr$ ratio is 0.7047 ± 4 (two sigma uncertainties). This age is interpreted as the intrusion age of the granite.

To the authors' knowledge there have been no previous isotopic studies on the age of mineralization in the Sanjia district. Our investigations involved Rb-Sr dating of hydrothermal vein sericite and K-feldspar from the Wangtoushan deposit, and dating of vein quartz from the Sanjia deposit by the $^{39}Ar/^{40}Ar$ incremental heating method. The Rb-Sr analytical data are given in Table 3.6 and the $^{39}Ar/^{40}Ar$ results are given in Fig. 3.13.

The Rb-Sr age of sericite collected from the Wangtoushan deposit (sample 6313) is 184 Ma. This is considered to be a reliable estimate of the mineralization age since the calculation is not sensitive to the initial $^{87}Sr/^{86}Sr$ ratio. The K-feldspar sample 6273 yields a concordant age of 179 Ma but a

Table 3.6. Rb and Sr isotopic data from hydrothermal minerals from the Wangtoushan deposit

Sample	Mineral	Rb	Sr	$^{87}Rb/^{86}Sr$	$^{87}Sr/^{86}Sr^a$	Age[b]
6313	Sericite	717	9.70	226	1.3005(1)	184 ± 2
6271	K-feldspar	618	312	5.75	0.71938(5)	115 ± 25
6273	K-feldspar	469	224	6.09	0.72550(4)	179 ± 25

[a] Numbers in parentheses give uncertainty of last digits.
[b] Age in Ma with uncertainty based on 1% relative error in $^{87}Rb/^{86}Sr$ and an initial $^{87}Sr/^{86}Sr$ ratio of 0.710.

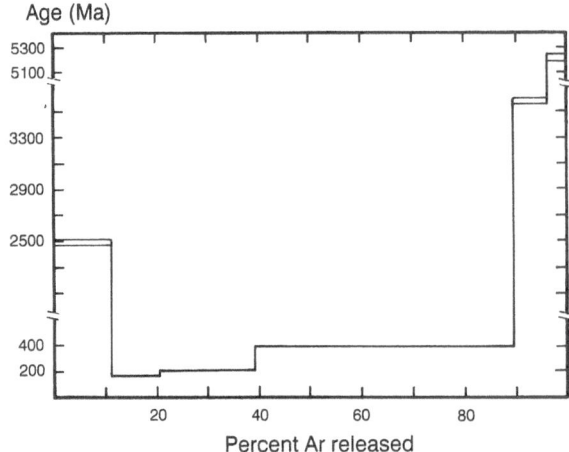

Fig. 3.13. Ar-Ar age-release curves of a sample of vein quartz from the Sanjia deposit. *Double lines* reflect analytical uncertainty. Analyses performed by the Ministry of Metallurgical Industry, Beijing

second sample from the same area (6271) gives a much younger age of 115 Ma. The difference in age between the two samples cannot be real since both came from neighboring veins in the same fracture system. It is concluded that the younger age represents loss of radiogenic ^{87}Sr from the sample.

These mineral ages are slightly older than that of the Sanyihe Granite, in apparent contradiction to the field relationships. This may not be a real discrepancy, since the hydrothermal minerals were taken from the Wangtoushan deposit some kilometers from the Sanyihe Granite, and the Wangtoushan Granite has not been dated. More dating should be done in the Sanjia district, but it suffices for the present purpose to note that gold mineralization and Yanshanian granites are nearly contemporaneous.

^{39}Ar/^{40}Ar Dating

The sample used for dating by the ^{39}Ar/^{40}Ar incremental heating method was vein quartz from the Sanjia deposit. The results shown in Fig. 3.13 indicate a plateau age of 168 ± 3 Ma, which is interpreted as the age of quartz crystallization in the vein. This agrees fairly well with the Rb-Sr ages from the Wangtoushan deposit, and it confirms a Mesozoic age for the mineralization in the district.

3.2.6 Fluid Inclusions

The data on fluid inclusions from the Sanjia district are based on analyses performed in the course of this project. The investigation involved microthermometric analyses of samples of fluorite and quartz from the Wangtoushan deposit, and chemical analyses of bulk fluid released by decrepitation from samples of vein quartz from the Wangtoushan and Sanjia deposits.

Microthermometry

Most vein quartz in the Sanjia district is clouded with minute secondary inclusions and therefore not suited for microthermometry. However, fluorite contains measureable inclusions which, based on their size, regular form, and random arrangement in the host, are thought to be primary. In one sample, quartz inclusions in fluorite were free enough of the secondary fluid inclusion "clouds" that measurements on primary inclusions (based on the criteria above) could be made. Table 3.7 gives a summary of microthermometric data from fluorite and quartz samples from the Wangtoushan deposit.

The inclusions in fluorite are large (10 to 25 microns in diameter) and regular in shape. All inclusions contain mixed CO_2-rich fluids, as evidenced by the formation of clathrates on cooling. Only rarely is liquid CO_2 visible in these inclusions at room temperature or on cooling. The degree of fill is 70 to 80%. The "primary" inclusions in quartz are rare, but the few examples

Table 3.7. Summary of microthermometric data from samples of fluorite and quartz from the Wangtoushan deposit

Host mineral	Sample	T_m CO_2	T_m clathrate	T_h CO_2	T_h final
Fluorite	6580a	−56.9	7.6	28.1	193
	(n = 10)	(x)	(0.3)	(1.9)	(12.8)
	6580b	−56.6	8.0	22.8	193
	(n = 14)	(x)	(0.1)	(2.1)	(20)
	6267a	−58.0	6.8	−5.0	335
	(n = 13)	(x)	(0.2)	(3.5)	(x)
Quartz	6580-2	−58.3	6.2	22.2	>290
	(n = 8)	(0.1)	(0.3)	(5.7)	−

All temperatures represent average values in °C.
T_m, melting temperature; T_h, homogenization temperature; n.d., not determined; (x), number of measurements too small for statistical analysis.
The number of measurements (n) and the standard deviation are given in parentheses.

found have properties similar to the inclusions in fluorite, and are probably of the same generation.

Salinities based on clathrate melting temperatures in both host minerals range from 6–8 wt% NaCl equivalent. The CO_2 melting temperatures indicate nearly pure CO_2 in some samples and minor impurity of the CO_2 phase in others. The final homogenization temperature of inclusions in both host minerals was difficult to determine because of decrepitation. In fluorite, a number of inclusions homogenized to the liquid phase in the range 180–240 °C, a few isolated inclusions homogenized at 330–345 °C and many inclusions decrepitated at about 290 °C. In quartz, most inclusions also decrepitated at about 290 °C. Therefore, a miminum of about 300 °C is taken as a reasonable estimate of the homogenization temperature. From the position of the CO_2-H_2O solvus at the appropriate salinity given by Brown and Lamb (1989), a minimum trapping pressure of about 2.5 kbar can be derived from these data. The true conditions of trapping can potentially be estimated using the fluid isochores combined with oxygen isotopic thermometry of vein quartz and K-feldspar. Unfortunately, the analyzed quartz and feldspar from the Wangtoushan deposit are not in isotopic equilibrium, and therefore no meaningful temperatures can be calculated.

Chemical Analyses of Fluid Contents

The average composition of bulk fluid released from inclusions in vein quartz of the Sanjia and Wangtoushan deposits is shown in Fig. 3.14. It must be emphasized that the fluids analyzed may represent a mixture of several generations, although the samples were decrepitated above 250 °C to minimize the contribution from low-temperature secondary fluids. The fluid composition in both deposits is very similar in terms of the dissolved cations. These are dominated by Na, with much less K and Ca, and very minor Mg. The CO_2 concentration is about 4 mol% from the Wangtoushan samples and only 1.6 mol% from Sanjia. Note that the lower CO_2 concentration in the Sanjia samples may simply reflect a larger component of aqueous secondary inclusions in these samples compared with those from Wangtoushan.

3.2.7 Stable Isotopic Data

This section presents oxygen and carbon isotopic data from quartz, K-feldspar, and carbonate gangue from ore veins in the Wangtoushan, Sanjia and Xinglonggou deposits. The main purpose of the analyses was to provide estimates of the temperature of mineralization, and only this aspect is discussed here. The implications of the data to the source of mineralizing fluids are discussed in Chapter 4.6.2.

Table 3.8 lists the analytical data. The isotopic composition of quartz falls in the range of $\delta^{18}O$ (SMOW) from 10.2 to 11.4‰, which is identical to vein

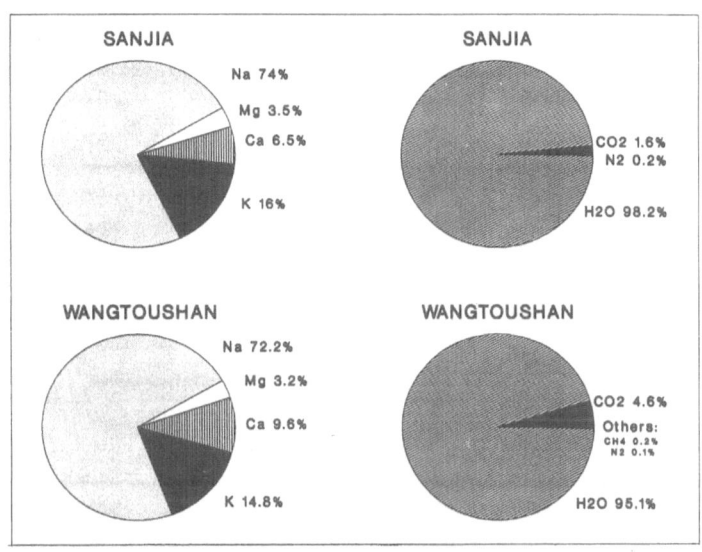

Fig. 3.14. Pie diagrams showing the average composition of bulk fluid inclusion contents from vein quartz in the Wangtoushan and Sanjia deposits. From unpublished analyses of the Ministry of Metallurgical Industry, Beijing

Table 3.8. Carbon and oxygen isotopic composition of carbonate, quartz, and K-feldspar from hydrothermal veins in deposits of the Sanjia district

Sample	Deposit	Mineral	$\delta^{13}C$ PDP	$\delta^{18}O$ SMOW
6271	Wangtoushan	K-feldspar	–	9.9
6271	Wangtoushan	Quartz	–	11.4
6273	Wangtoushan	K-feldspar	–	11.2
6273	Wangtoushan	Quartz	–	10.2
6309	Wangtoushan	K-feldspar	–	11.3
6309	Wangtoushan	Quartz	–	10.6
6275	Wangtoushan	Dolomite	−6.7	13.5
6309	Wangtoushan	Dolomite	−4.8	13.6
6285	Xinglonggou	Dolomite	−2.7	8.3
6289	Xinglonggou	Dolomite	−6.8	13.2
6210	Sanjia	Dolomite	−5.5	11.4

The precision of analyses is better than ±0.2‰.

quartz from the Niuxinshan deposit (Table 3.4). However, the composition of K-feldspar samples are anomalous in that, in three out of four cases, they are richer in ^{18}O than coexisting quartz. This is contrary to the equilibrium fractionation trend (O'Neil 1986), and therefore the data cannot be used for thermometry and no temperatures are indicated in Table 3.8.

The carbon and oxygen isotopic compositions of carbonate minerals define a narrow range. The interpretation of these compositions is deferred to Chapter 4.6.2, where the data from all deposits are discussed together.

3.3 Yuerya District

The Yuerya gold district is located about 200 km northeast of Beijing, in Kuancheng county, eastern Hebei province. The location is shown in Fig. 3.1. Strictly speaking, only one gold deposit, the Yuerya deposit, occurs in the mining district but smaller deposits of the same geologic type occur within a few tens of kilometers to the northeast of the Yuerya deposit. These include the Baizhangzi, Yangzhangzi, and Maojiadan gold deposits. Only the Yuerya deposit is described in this text. The Yuerya deposit serves as the type deposit for the Yuerya type of gold deposits, i.e., those hosted by Mesozoic granites, discussed by Yu and Jia (1989). Wei (1989) gave a brief description of the Yuerya deposit.

After its discovery in 1887 the Yuerya gold deposit was operated on a small scale until 1965, when systematic prospecting and exploration work by the Ministry of Metallurgical Industry was completed. The present mine plant with an ore dressing and smelting facility was finished in 1968. The mine is operated by the state government.

3.3.1 Host Rock Lithology

The geology of the Yuerya deposit is shown on a simplified geologic map in Fig. 3.15. The host rocks consist of Middle Proterozoic very low-grade metamorphic carbonate rocks intruded by a Yanshanian granite stock and associated mafic and felsic dikes.

3.3.1.1 Proterozoic Rocks

The Proterozoic rocks in the Yuerya area comprise a marine sedimentary sequence of limestones, dolomites, and minor shales which belong to the upper part of the Gaoyuzhuang Formation, the uppermost formation of the Changcheng System. The lithologic sequence at Yuerya comprises, in ascending order, shales, interbedded shales and limestones, manganiferous dolomites, thickly bedded limestones, calcretes, and thinly bedded limestones. The entire sequence strikes NE–SW and dips 50°–70° to the NW. Minor folds and bedding-parallel faults occur, as described more fully in Chapter 3.3.2, but the sequence in general has a monoclinal form. The maximum age for the Gaoyuzhuang Formation is 1678 Ma based on isotopic dating of the underlying formation. The only isotopic constraint on the

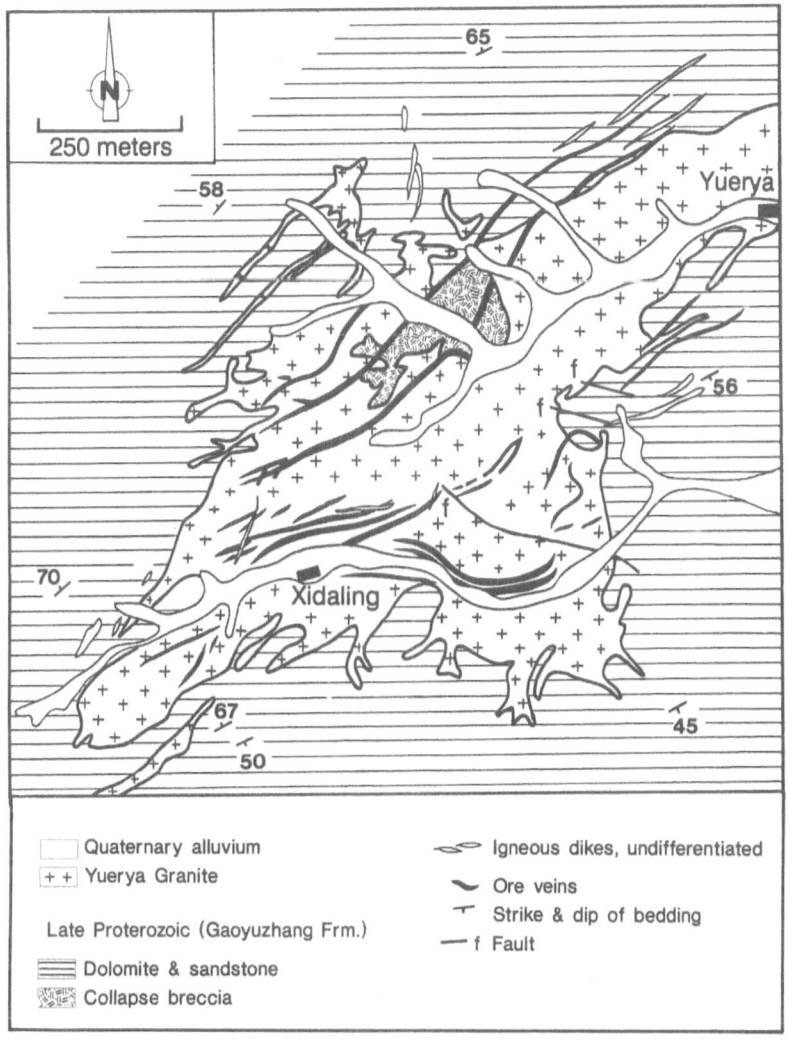

Fig. 3.15. Simplified geologic map of the Yuerya gold deposit

minimum age is a 1200 Ma age from the top of the succeeding Jixian system (see Chap. 2.2.2).

3.3.1.2 Granite and Dikes

The Yuerya Granite is about 2 km long and 700 m wide in outcrop extent, and is elongated NE–SW parallel to the regional strike of the country rocks (Fig. 3.15). The intrusion is highly irregular in form, with numerous apophyses of granite along bedding-parallel fractures in the country rocks. Many country rock inclusions are present in the granite, and in the western part of the intrusion a collapse breccia of roof fragments occurs. Previous radiometric dating has not given an unequivocal age for the granite. Yu and Jia (1989) give K-Ar ages of 149 and 169 Ma. Recent unpublished K-Ar determinations by the Ministry of Metallurgical Industry yielded three ages of 168 ± 3, 166 ± 3, and 163 ± 2 Ma. It seems therefore most likely that the age of the granite is around 165–170 Ma.

The Yuerya Granite consists of two intrusive phases. The earlier phase, and the only one exposed at the surface, is a light-colored medium- to fine-grained biotite granite. A later intrusive phase, which is known from mine workings and drill core, consists of reddish medium- to coarse-grained biotite granite. The mineral compositions of both granite types are similar, with quartz, K-feldspar, oligoclase, biotite, and apatite; however, the red granite contains less biotite than the light granite. The chemical compositions of the two granite phases are also very similar. The red granite is richer in SiO_2 (73–75 wt% vs 70–73 wt%), and slightly poorer in Fe, Mg, Ti, and Ca than the light granite, although in both granite types the sum of total Fe as FeO, MgO, CaO, and TiO_2 is generally less than 3 wt%. Total alkali concentrations are near 8%. The weight ratio of K_2O/Na_2O is near unity in fresh samples but rises rapidly with alteration. The moderate concentrations of Rb (150 ppm) and the relatively high Sr (150 ppm) and Ba (450 ppm) suggest a low degree of differentiation. Analytical data are summarized in Appendix 1.

Three groups of igneous dikes intrude both the granite and the country rocks along fracture zones striking NE–SW and NW–SE. The earliest dikes are granitic dikes of aplitic or rhyolitic texture which may be cogenetic with the Yuerya Granite. The granitic dikes are cut by ore veins. A second group of dikes, which postdates the ore vein formation, consists of porphyritic diorite. Lamprophyre dikes also occur locally and these, too, postdate mineralization. No radiometric age data are available from the dikes.

3.3.2 Host Rock Structures

As described above, the Late Proterozoic sequence has a monoclinal form striking 30°–35° and dipping 50° to 70° NW. Minor folds with NE-striking axes are locally present, one example being a small syncline near the southeast margin of the granite. Apart from minor folds and flexures in the Proterozoic rocks, the deformation in the district is dominated by faults and fractures.

The Yuerya deposit and the other gold deposits of similar type (Baizhangzi, Yangzhangzi, and Maojiadan) are located along the southern portion of the regional Tangdaohe-Lingyuan deep fracture zone which was described in Chapter 2.3.3 and shown in Figs. 2.5 and 2.6. Two major groups of faults and fracture zones are recognized within the mining district. The first group is related to the regional Tangdaohe-Lingyuan fracture zone and comprises reverse and oblique-slip faults which strike NE–SW (55°–65°) and dip 30°–70° to the NW. These faults are present both within the Yuerya Granite and in the Proterozoic sequence. The mineralized veins and alteration zones generally occur along these NE-trending structures. A second group of faults and fracture zones strikes WNW 280°–290° and dip 5°–30° to the NE; these are not significant hosts for mineralization, and usually show evidence of postmineralization movement.

3.3.3 Gold Mineralization

The gold mineralization in the Yuerya deposit is almost exclusively confined to quartz veins and/or disseminated alteration zones within the Yuerya Granite. In 1989, 20 ore-bearing vein zones were known. The veins and associated alteration zones strike NE–SW and dip moderately to the NW. The greatest concentration of veins and the richest mineralization occurs in the southern and eastern parts of the Yuerya intrusion. Most of the ore occurs within the light granite near its (underground) contact with the red granite; a second favorable site for mineralization is at the contact of the granite with the country rocks. Although the fractures which host the veins continue from the granite into the carbonate rocks, mineralization rarely extends more than 50 m out from the granite contact.

3.3.3.1 Form of the Ore Bodies

The ore bodies in the Yuerya deposit may be subdivided into two types based on their form, namely vein-type and disseminated type. The vein-type ore bodies occur both within the granite and at the granite-country rock contact. Within the granite the veins are typically 30–50 cm thick and the ore contains up to 100 g/t gold, on average 50 g/t. Bordering the veins are relatively narrow selvages of altered rock with disseminated mineralization. Thinner quartz veinlets with lower grade often occur in secondary fractures associated with the main veins. At the granite-country rock contact the vein zones are characterized by more intense shearing than within the granite. Ore bodies in this setting are irregular lenses of sheared sulfide-rich quartz veins. The thickness of the ore bodies and the extent of alteration and mineralization are greatest in extensional "flats", where the dip of the host fractures abruptly changes. This ore type contains up to 50 g/t gold. Where the ore veins extend into the country rocks the grade falls rapidly. Individual mineralized zones are typically between 100 and 900 m long (average 200 m) and up to 5–10 m wide (average 0.3–1 m). Present data show mineralization persisting to 300–400 m down dip.

The disseminated type of mineralization is found only within the granite. The disseminated ore bodies are developed around networks of fine secondary fractures associated with the main NE–SW-trending fracture zones. In this type of ore there is a continuous transition between economic ore bodies and the wall rock. In general the economic ore bodies are about 1–2 m thick (up to 10 m), whereas the entire mineralized zone is commonly more than 10 m thick. Gold concentration averages 5–10 g/t and ranges to a maximum of 30–50 g/t.

3.3.3.2 Macroscopic Description of the Ores

The vein-type ores consist of sheared or brecciated quartz with dominantly pyrite and occasionally pyrrhotite as the main sulfide mineral. The richest ores consist of well over 50 vol% sulfides. Galena and sphalerite are commonly developed on the margins of the ore veins. Rare veinlets and massive patches of sulfide minerals with little or no quartz may occur in the altered wall rocks adjacent to the quartz veins. These consist of intergrown pyrite, chalcopyrite, galena, and sphalerite, with accessory tetrahedrite and rare calaverite. The gangue minerals of the vein-type ore in granite are mainly quartz, sericite, calcite, and minor barite. The calcite is generally found in late cross-cutting veinlets. Where the mineralization extends into the country rocks the above-mentioned gangue minerals may be joined by chlorite.

The disseminated ores consist of pervasively silicified and sericitized granite which is permeated by thin quartz and sulfide veinlets (10–30 cm thick, locally up to 10 cm thick). Pyrite occurs as scattered grains in the rock, or as concentrated centimeter-sized nests and patches. Gangue minerals in the disseminated type of ore include quartz, sericite, albite, carbonate, and kaolinite. The carbonatization and kaolinization postdate mineralization.

3.3.3.3 Ore Petrography and Paragenesis

The following ore minerals have been found in the Yuerya deposit: azurite, bismuthinite, bornite, calaverite, chalcocite, chalcopyrite, galena, hessite, malachite, molybdenite, native gold, pyrite, pyrrhotite, scheelite, sphalerite, tennantite, and tetrahedrite. The paragenetic sequence of mineralization at Yuerya is complex. The mineralization can be divided into four main phases from early to late as follows:

a) pyrite-pyrrhotite ± quartz phase,
b) quartz-pyrite phase,
c) quartz-polymetallic sulfide phase,
d) telluride-carbonate-sulfosalt phase,
e) secondary sulfide stage.

Detailed petrographic descriptions are available only for the occurrence of native gold. Most of the gold formed in the second and third phases of mineralization (b and c above) where gold is mainly associated with pyrite. Minor amounts of gold also formed later in the paragenetic sequence in phase d together with telluride minerals. Gold has been observed in the following settings:

a) at the rims or grain boundaries of pyrite,
b) associated with sphalerite, quartz and galena in fracture-fillings in the early stage pyrite,
c) associated with telluride (hessite) aggregates,
d) along grain boundaries or in cracks of fine-grained vein quartz or, rarely, feldspar,
e) with secondary chalcocite filling cracks in late-stage pyrite.

The native gold grains are generally between 0.1 and 0.4 mm in diameter. Grains up to 1 mm in diameter and stringers of gold up to 3 mm long occur as well. Microprobe analyses of gold grains showed 23 wt% Ag (Yu et al. 1989).

3.3.4 Wall Rock Alteration

Alteration of Carbonate Rocks
In the limestone/dolomitic country rocks of Yuerya Granite the effects of alteration due to the mineralized quartz veins are superimposed on an earlier contact metamorphism. The contact metamorphism produced recrystallization of limestone, the development of hornfels in the shaley horizons, and local zones of skarn minerals (diopside, tremolite, ± grossular). The alteration associated with mineralization involved intense silicification, lesser chloritization and sericitization, and minor pyritization.

Alteration of Granite
Wall rock alteration within the granite is associated with both the vein type and the disseminated type of mineralization. The alteration mineralogy in both types is essentially the same, the only difference being that the latter affects a greater volume of rock, and that in the disseminated type of ore, the altered rock is mineralized to ore grade. The alteration involves pyritization, sericitization, silicification, and albitization, with minor chloritization and kaolinization. Of these, the silicification, pyritization, sericitization, and albitization are closely related to the gold mineralization; carbonate and kaolin generally formed later.

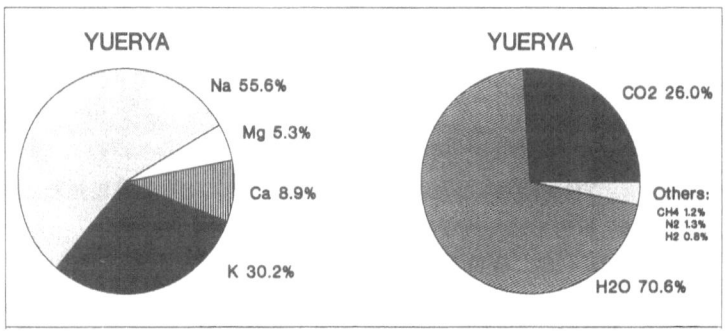

Fig. 3.16. Pie diagrams showing the average composition of bulk fluid inclusion contents from vein quartz in the Yuerya deposit. Source of data: Yu and Jia (1989) and Yu et al. (1989)

3.3.5 Age of Mineralization

The mineralization of the Yuerya deposit clearly cross-cuts the Yuerya Granite and must therefore be younger than that intrusion. The age of the Yuerya Granite is a matter of some controversy as discussed above, but the most likely age is considered to be 165–170 Ma. The only isotopic age data obtained from hydrothermal minerals at Yuerya is a K-Ar age of 200 Ma from "fine scaly sericite" obtained by Yu and Jia (1989). Either this age is too old, for example because of excess Ar, or the granite ages are too young. More work is needed to clear up this discrepancy.

3.3.6 Fluid Inclusions

Data from studies of fluid inclusions in quartz from the Yuerya deposit are reported in Yu and Jia (1989) and Yu et al. (1989). The reports include total homogenization and decrepitation temperatures and chemical analyses of bulk inclusion contents. According to Yu and Jia (1989), the final homogenization temperatures of inclusions in three samples range from 259–390 °C, with homogenization to the liquid phase. The maximum decrepitation temperature peak for the same samples ranges from 250–273 °C. Yu et al. (1989) reported a higher range of decrepitation temperatures (291–345 °C) from ten analyses.

Figure 3.16 shows the average chemical composition of bulk inclusion fluids plotted from the data given by the above-mentioned authors. The results show that Na forms only about half of the dissolved cations, the remainder being mostly K with lesser Ca and very minor Mg. Carbon dioxide makes up 26 mol% of the fluid, and other gases combined constitute about 3 mol%.

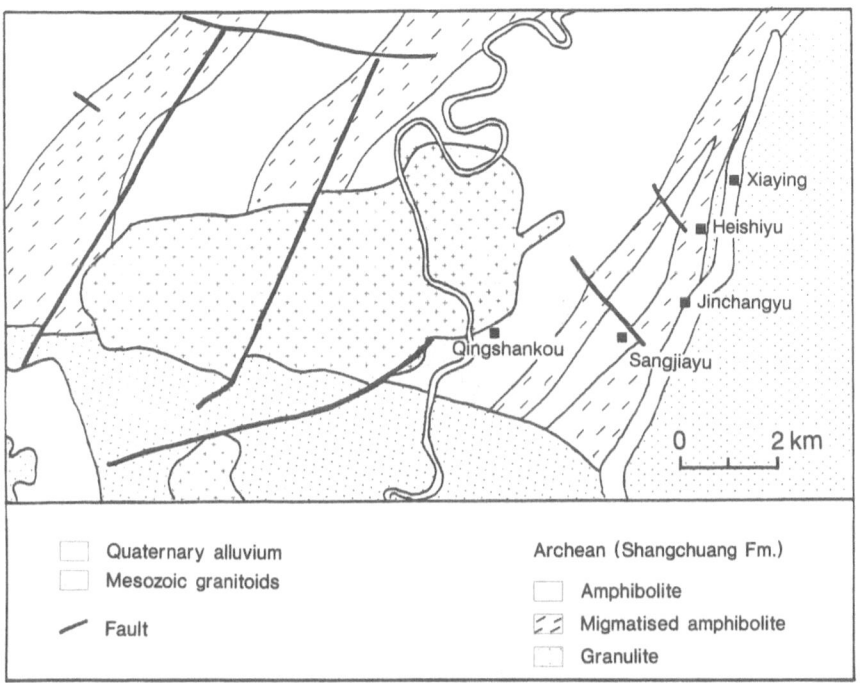

Fig. 3.17. Simplified geologic map of the Jinchangyu district

3.4 Jinchangyu District

The Jinchangyu deposit is located about 150 km northeast of Beijing in Kuancheng county, Hebei province. (see Fig. 3.1). It is the largest gold deposit in eastern Hebei province and one of the most important in China. Gold was mined at Jinchangyu during the Tang Dynasty (600–900 A.D.) and it was one of the three largest gold producers in China during the Qing Dynasty (1644–1911 A.D.). After the founding of the People's Republic, the deposit was systematically reexamined, beginning in 1963, and production renewed in 1968. The Jinchangyu mine is operated by the state government with entirely underground workings. The Jinchangyu deposit covers an area of about 6 km long and up to 900 m wide. It is divided into three geographic sections from north to south: Heishiyu section, Jinchangyu section, and Sanjiayu section (see Fig. 3.17). Some authors refer to each section as a separate deposit; however, in this text all are treated together. Figure 3.18 shows a simplified geologic map of the deposit.

The geology of the Jinchangyu deposit has been reviewed briefly by Gao (1986), Sang and Ho (1987), L.S. Yang (1988), and in more detail by Yu and Jia (1989). It serves as the type deposit for the Jinchangyu type of metamorphic-hosted gold deposits proposed by Zhu (1989).

103

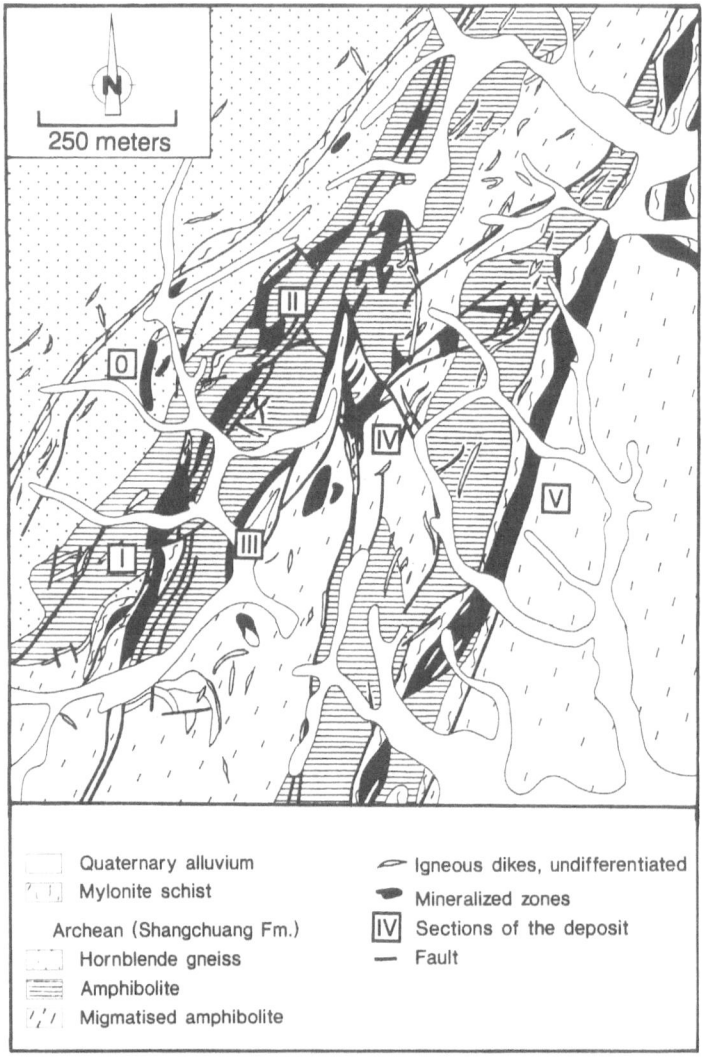

Legend:

- Quaternary alluvium
- Mylonite schist

Archean (Shangchuang Fm.)
- Hornblende gneiss
- Amphibolite
- Migmatised amphibolite

- Igneous dikes, undifferentiated
- Mineralized zones
- IV Sections of the deposit
- Fault

Fig. 3.18. Simplified geologic map of the Jinchangyu deposit

3.4.1 Host Rock Lithology

The dominant country rocks in the Jinchangyu district are Archean amphibolites, gneisses and granulites of the Shangchuang Formation, Qianxi Group. Igneous rocks are represented by a Yanshanian pluton of biotite granite (Qingshankou Granite) and by numerous dikes of various composition whose age is also presumed to be Yanshanian.

104

3.4.1.1 Archean Rocks

The Archean host rocks in the vicinity of Jinchangyu belong exclusively to the Shangchuang Formation of the Qianxi Group and contain the following main rock types:

- migmatized amphibolites, locally containing biotite or garnet,
- magnetite-quartz amphibolites grading into magnetite quartzites,
- two-pyroxene granulites with layers of amphibolite, the latter probably due to retrograde metamorphism (K.Y. Wang et al. 1985; Jahn et al. 1987),
- hornblende-plagioclase orthogneisses (metatonalites).

The main type of country rock in the vicinity of the Jinchangyu deposit is migmatized amphibolite which contains local magnetite-quartz rich layers. Minor amounts of garnet amphibolite and biotite amphibolite also occur. Migmatization in the Shangchuang Formation is intense in the Jinchangyu area (Fig. 3.17). The migmatization produced K-feldspar blastesis and abundant veins and schlieren of leucosome material. Retrograde metamorphism is commonly shown by replacement of plagioclase by albite and sericite, and the replacement of pyroxene and hornblende by fibrous amphibole, biotite, and chlorite. In the vicinity of some fault zones. the amphibolite and granulite and shered and altered to chlorite-sericite schists.

3.4.1.2 Granite and Dikes

There is no granite intrusion directly within the Jinchangyu mining district. The nearest granite, the Qingshankou Granite, crops out about 3 km to the west (see Fig. 3.17). The Qingshankou Granite is a multiple intrusion of Early Yanshanian age (K-Ar age 196 Ma according to Yu and Jia 1989). The early phase of the intrusion consists of medium to coarse-grained biotite-hornblende quartz diorite. This was followed by the main intrusive phase of coarse-grained biotite granite consisting of K-feldspar (50%), oligoclase (20%), quartz (25%), and biotite (5%), with accessory muscovite and zircon. The latest phase of the pluton is made up of leucocratic alkali-feldspar granite with accessory muscovite. The chemical composition of the Qingshankou Granite (main phase, biotite granite) is characterized by SiO_2 concentrations of 73 wt%, total alkalis near 9 wt%, (K_2O/Na_2O weight ratio near 1), and low concentrations of Fe, Mg, Ti, and Ca (the sum of total Fe as FeO, MgO, TiO_2, and CaO is less than 3 wt%). The moderate concentration of Rb (150 ppm) and the relatively high Sr (120 ppm) and Ba (450–600 ppm) suggest a low degree of differentiation. The chondrite-normalized REE distribution pattern of the granite (not shown) has a smooth decrease in enrichment from the light to the heavy REE and a weak negative Eu anomaly. The composition is summarized in Appendix 1.

Several groups of igneous dikes occur in the district (not differentiated in Fig. 3.18). The most common dikes consist of granitic porphyry. Other dikes include albitic aplite, trachyte, diorite porphyry, and hornblende-phyric lamprophyre. All of the dikes strike NE–SW to NNE–SSW, parallel to the main fault and fracture directions in the mining district. Field relations do not reveal the relative timing of all types of dikes with mineralization; however, the granitic porphyry dikes clearly predate mineralization, and the lamprophyres are clearly later.

3.4.2 Host Rock Structures

The structures of the basement rocks at the Jinchangyu district are characterized by at least two periods of folding and by later ductile and brittle faulting. The Qingshankou Granite and igneous dikes postdate all folding episodes and show only brittle deformation.

3.4.2.1 Folds and Foliation

The main trend of foliation in the Archean rocks strikes NE–SW (40°–60°) and dips to the NW at 40°–70°. This foliation is axial planar to tight folds which L.S. Yang (1988) attributed to the Fuping Orogeny at about 2500 Ma. According to L.S. Yang (1988), these NE–SW-directed folds overprint earlier Archean folds with E–W-trending axes. Evidence of later E–W-trending open folds is lacking at Jinchangyu although such folds are present regionally (see Fig. 2.5).

3.4.2.2 Faults and Fractures

The structural style of the Jinchangyu district is dominated by a complex series of faults and fracture zones which include both Early Proterozoic and Mesozoic elements. The Early Proterozoic faults are oriented NNE to NE, and they consist of chlorite and sericite-rich "mylonite schist" which, according to L.S. Yang (1988), formed during the Fuping Orogeny. The nature of these Precambrian shear zones is obscured by the fact that they have been reactivated by the Yanshanian deformation, and intruded by Mesozoic dikes and quartz veins. The Early Proterozoic fault zones apparently formed under conditions of N–S-directed stress, which produced conjugate NE–SW and NW–SW-trending shear zones, E–W-striking reverse faults, and N–S-striking normal faults. The Yanshan Orogeny produced intense and widepread faulting at Jinchangyu which reactivated and partly overprinted the older structures. The Yanshanian compressive structures (including reverse and strike-slip components) trend NNE–SSW, and associated tension fractures are oriented WNW–ESE. These later faults offset the early mylonite-schist zones and many of the mineralized structures as well (Fig. 3.18).

3.4.3 Gold Mineralization

Gold mineralization in the Jinchangyu ore deposit is found in steeply dipping NE–SW-trending shear zones on the NW limb of the Jinchangyu anticline. The Jinchangyu deposit is divided into six sections of vein zones numbered O-V from west to east (Fig. 3.18). Each section is truncated by cross-faults. Present workings and drill hole data indicate mineralization to a depth of at least 500 m.

3.4.3.1 Form of the Ore Bodies

There are a total of 17 producing ore bodies in the Jinchangyu deposit (as of 1989). The ore bodies are 50–150 m long (maximum 300 m) and 1–6 m thick (maximum over 30 m). In general, the ore bodies are in the form of quartz or quartz-albite veins within sheared fault zones (shown in black in Fig. 3.18) although mineralization is also found disseminated in intensely sheared and altered host rocks. The host shear zones strike NE–SW and dip 70°–80° variably NW or SE. According to L.S. Yang (1988), the shear zones show evidence of mylonitization under greenschist-facies conditions, i.e., they are synmetamorphic, and the gold-bearing quartz or quartz-albite veins were introduced later. The veins themselves show evidence of both pre- and postmetamorphic phases of mineralization. Premetamorphic, barren, or subeconomic veins are concordant with foliation of the host rocks and may show ptygmatic folding, whereas the economic veins are discordant, unfolded, and clearly postmetamorphic.

The host shear zones have been divided into three groups according to their strike direction. These are designated simply as the 0°, 20°, and 60° groups. The 20° group is the most intensely mineralized and an estimated 60% of the ore in the deposit is mined from this group of structures. The areas of intersection of two or more shear zones are particularly favorable for mineralization.

3.4.3.2 Macroscopic Description of the Ores

The ore bodies in the Jinchangyu deposit show evidence of several generations of movement and refilling by quartz and other gangue minerals. Three stages are recognized (unpublished reports, Ministry of Metallurgical Industry). The first two stages, involving quartz and quartz-albite filling,

107

were poorly mineralized, and in some cases these feldspathic veins are clearly metamorphic, being folded and concordant with the host rock foliation. The main stage of mineralization is characterized by gray to white quartz gangue lacking albite, and veins of this generation are typically discordant, although some of the earlier folded veinlets were also enriched by the third-stage mineralization. The latest veining is by carbonate, which postdates mineralization.

There are systematic differences in the mineral assemblage of veins with different orientations. Thus, the 0° group of faults contains simple, weakly mineralized quartz veins with little wall rock alteration, the 20° group contains rich veins showing multiple stages of mineralization and intense wall rock alteration, and the 60° group is characterized by early-stage quartz-albite veins with chloritized wall rocks and only minor mineralization. The occurrence of sulfide minerals in the veins may be disseminated, concentrated in vein-like seams within the quartz (reopening structures), or along the vein walls. In the richest ore the sulfides have a massive structure. Sulfide minerals typically make up about 10 vol% of the ore-bearing veins of the 20° group, and correspondingly less in the other vein groups.

3.4.3.3 Ore Petrography and Paragenesis

The following ore minerals were found in the Jinchangyu deposit: argentite, bornite, calaverite, chalcocite, chalcopyrite, galena, hessite, magnetite, malachite, molybdenite, native gold, pyrite, pyrrhotite, and sphalerite. Despite the importance of this deposit, no systematic studies of ore paragenesis have been published. The only detailed petrographic descriptions are of pyrite and gold, taken from unpublished reports of the Ministry of Metallurgical Industry.

Pyrite
Pyrite is the most abundant ore mineral and it occurs both early and late, and in both wall rocks and quartz veins. The early generation of pyrite is characteristically euhedral in shape with a grain size of 3–5 mm. Later pyrite forms fine-grained anhedral aggregates in stringers filling cracks in quartz, early pyrite, and other gangue minerals. The richest ore is associated with this generation of fine-grained anhedral pyrite. The latest pyrite generation is associated with galena, chalcopyrite, and bornite in cracks of early-stage pyrite.

Gold
Gold occurs most typically associated with fine-grained anhedral pyrite of the second generation mentioned above. The gold occurs as fracture fillings, as replacements on pyrite rims and grain boundaries, and as inclusions in pyrite. Associated minerals are chalcopyrite, calaverite, and argentite.

Pyrite-associated gold makes up about two-thirds of the gold in the ores. The remaining third occurs within quartz. The grain size of gold ranges from about 5 to 20 microns. The silver concentrations in gold range from about 5 to 28 wt%.

3.4.4 Wall Rock Alteraton

Wall rock alteration of the metamorphic host rocks at Jinchangyu is ubiquitous. The width of the alteration zones varies according to the size of the associated vein and, especially, on the degree of shearing of the host rocks. The simple quartz veins of the 0° group, as mentioned above, have developed weak alteration zones with only a few decimeters to meters thickness whereas the richly mineralized and complex veins of the 20° group are associated with intense alteration of the host rocks, with sericitization zones reaching 40 m in thickness. The selvages of altered rock around the quartz veins typically show a systematic zonation of mineral assemblage, with an inner zone of pyritization, silicification, and sericitization followed outward by zones of chloritization with carbonatization, and of weak chloritization alone. Pyritization, sericitization, and silicification are most closely related to the gold mineralization. Carbonatization is generally later than the other alteration phases.

3.4.5 Age of Mineralization

Field evidence for an early, pre-metamorphic stage of mineralization in the Jinchangyu deposit includes folded, concordant quartz-albite veins, and greenschist-facies ductile shear zones which are mineralized to subeconomic grades. On the other hand, there is abundant evidence to show that the main stage of mineralization occurred much later. Most of the economic ore bodies show brittle deformation and occupy discordant structures. Undeformed, discordant granitic dikes are commonly cut by mineralized veins, and these dikes are interpreted to be of Yanshanian age, perhaps related to the nearby Qingshankou Granite, whose isotopic age is 196 Ma. No age dating of the igneous dikes has yet been made.
Published ages of hydrothermal sericite from the Jinchangyu deposit support the view that the mineralization is Mesozoic in age. Yu and Jia (1989) reported two sericite K-Ar ages of 170 and 155 Ma, and Yu et al. (1989) reported a K-Ar age of 192 Ma of sericite from a hydrothermally altered (unmineralized) fault zone. Sang and Ho (1987) cited a further K-Ar age of 197 Ma from unspecified material of wallrock alteration. Further evidence of a Mesozoic age for the Jinchangyu mineralization is a Pb-Pb model age of 133 Ma cited by Yu and Jia (1989) (see Chap. 4.2.3). The agreement among these ages is not close, but they suffice to confirm the field evidence that the main stage of mineralization was Mesozoic in age.

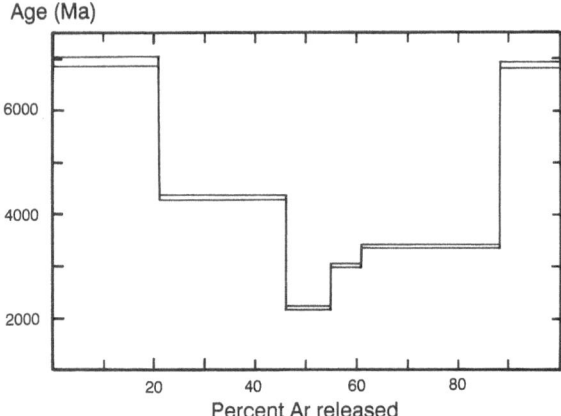

Age (Ma)

Percent Ar released

Fig. 3.19. Ar-Ar age-release curves from a sample of vein quartz from the Jinchangyu deposit. *Double lines* reflect analytical uncertainty. Analyses performed by the Ministry of Metallurgical Industry, Beijing

$^{39}Ar/^{40}Ar$ Dating

The results of $^{39}Ar/^{40}Ar$ dating of vein quartz from the Jinchangyu deposit by the incremental heating method are shown in Fig. 3.19. The analyses were performed by the Ministry of Metallurgical Industry, Beijing. The results show a highly disturbed Ar system with no evidence of Mesozoic age (compare the results from the Niuxinshan, Sanjia deposits in Figs. 3.7 and 3.13. This result is geologically unrealistic because the sample was taken from a quartz vein which cross-cuts Yanshanian dikes and must therefore be Mesozoic. The early age probably reflects Ar derived from the wall rocks and trapped in quartz as mineral or fluid inclusions.

3.4.6 Fluid Inclusions

Data from studies of fluid inclusions in quartz from the Jinchangyu deposit are reported in Yu and Jia (1989), Sun et al. (1989), and Yu et al. (1989). These authors give chemical analyses of bulk fluid contents and some microthermometric data.

According to Yu and Jia (1989), the final homogenization temperatures of inclusions in 17 samples range from 256–370°C, with homogenization to the liquid phase or critical. The decrepitation temperatures for the same samples range from 134–325°C. Yu et al (1989) reported decrepitation temperatures ranging from 140–365°C (mean 319°C) from 14 samples.

The average chemical composition of inclusion contents from Yu and Jia (1989) and Yu et al. (1989) is replotted in Fig. 3.20. Sodium constitutes only about half of the dissolved cations, the rest being mainly Ca and K, with

110

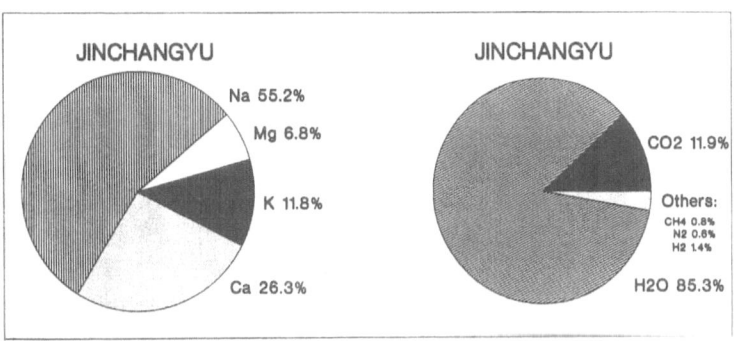

Fig. 3.20. Pie diagrams showing the average composition of bulk fluid inclusion contents from vein quartz in the Jinchangyu deposit. Analytical data from Yu and Jia (1989) and Yu et al. (1989)

minor Mg. Carbon dioxide makes up about 12 mol% of the fluid, the other dissolved gases together constitute about 3 mol%.

3.5 Banbishan District

The Banbishan gold district is located near the Qinglong River in eastern Qinglong county, Hebei province (Fig. 3.1). Mining at Banbishan began in 1982, and in 1985 the operation was taken over by the Qinglong county government. The gold district occurs in a 12-km-long zone, elongate NE–SW, which includes several gold occurrences. Apart from the Banbishan deposit, the following gold deposits and prospects occur in the area: Dayuzhangzi, Miaozhangzi, Zhangzhangzi, Wangzhangzi, Heluobao, Dakuaidi, and Shajinggou. The deposits are small in scale and many are subeconomic. Placer mining is also done on a small scale in the area along the Qinglong River. The following description concerns only the Banbishan deposit. A simplified geologic map of the deposit is shown in Fig. 3.21.

3.5.1 Host Rock Lithology

The host rocks of the Banbishan gold deposit mainly consist of a series of low- to medium-grade metamorphosed volcano-sedimentary rocks belonging to the Early Proterozoic Zhuzhangzi Group. These are cut by mafic and felsic igneous dikes of presumed Yanshanian age (Fig. 3.21).

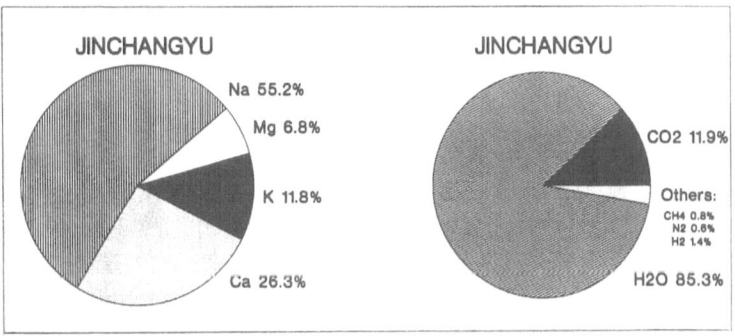

Fig. 3.20. Pie diagrams showing the average composition of bulk fluid inclusion contents from vein quartz in the Jinchangyu deposit. Analytical data from Yu and Jia (1989) and Yu et al. (1989)

minor Mg. Carbon dioxide makes up about 12 mol% of the fluid, the other dissolved gases together constitute about 3 mol%.

3.5 Banbishan District

The Banbishan gold district is located near the Qinglong River in eastern Qinglong county, Hebei province (Fig. 3.1). Mining at Banbishan began in 1982, and in 1985 the operation was taken over by the Qinglong county government. The gold district occurs in a 12-km-long zone, elongate NE–SW, which includes several gold occurrences. Apart from the Banbishan deposit, the following gold deposits and prospects occur in the area: Dayuzhangzi, Miaozhangzi, Zhangzhangzi, Wangzhangzi, Heluobao, Dakuaidi, and Shajinggou. The deposits are small in scale and many are subeconomic. Placer mining is also done on a small scale in the area along the Qinglong River. The following description concerns only the Banbishan deposit. A simplified geologic map of the deposit is shown in Fig. 3.21.

3.5.1 Host Rock Lithology

The host rocks of the Banbishan gold deposit mainly consist of a series of low- to medium-grade metamorphosed volcano-sedimentary rocks belonging to the Early Proterozoic Zhuzhangzi Group. These are cut by mafic and felsic igneous dikes of presumed Yanshanian age (Fig. 3.21).

Luzhangzi Formation

The lower part of the formation consists of orthoamphibolites with relict pillow structures, amygdales, vesicles, and doleritic texture. Overlying the amphibolites is a unit of hornblende-biotite schists and leptites whose protoliths were probably andesitic tuffs or lavas.

Zhangjiagou Formation

This formation contains metamorphosed agglomerate lavas intercalated with metaconglomerates and overlain by biotite leptites. The metaconglomerates consist of a biotite leptite matrix with stretched clasts of intermediate-felsic volcanic rocks and minor vein quartz. The conglomerates grade upward into biotite leptites, and they also grade laterally by a change of facies into the agglomerate lavas.

Shangbaichengzi Formation

This formation is the main host to ore in the Banbishan deposit. The rocks consist of weakly banded hornblende-biotite leptites in the lower unit succeeded by biotite-muscovite leptites. The protoliths are thought to be graywacke-type sedimentary rocks. Relict cross-bedding has been observed in parts of the lower unit.

Zhalanzhangzi Formation

This formation contains very fine-grained biotite schists and leptites.

3.5.1.2 Granites and Dikes

The nearest Yanshanian granite intrusion is the Louzishan Granite, located about 8 km SSE of Banbishan (not shown on Fig. 3.21). Within the mining district, igneous rocks are represented by four compositional groups of dikes. None of the dikes has been isotopically dated, but they are assumed, from structural evidence, to be of Yanshanian age. The earliest dikes consist of porphyritic aplite or rhyolite. These dikes strike NE–SW and dip about 50° to the NW. They intruded along bedding parallel fracture and fault zones in the Proterozoic rocks. The most important of the granitic dikes extends across the entire area of Fig. 3.21 and is up to 500 m in outcrop width. The granitic dikes are cut and mineralized by the ore-bearing veins. Rare dikes of porphyritic diorite cut the granitic dikes and are in turn cut by the later dike groups. No cross-cutting relationships with ore veins were observed. Lamprophyre dikes in the district are several tens of meters long, 2–5 m wide, and have variable strikes from nearly E–W to NNE–SSW, with moderate dips 40°–60° to the N and NW. No cross-cutting relationships were found between lamprophyre and mineralized veins. The latest and most common group of dikes in the district consists of equigranular diorite. The dikes intrude NE–SW-striking faults and fracture zones dipping 60°–80°

to the NW. The diorite dikes cut across mineralized quartz veins and shear zones, and therefore postdate mineralization.

3.5.2 Host Rock Structures

The two major structural units in the Banbishan district are the NE–SW-trending Qinglonghe Anticlinorium, which is a late-stage (Proterozoic) fold zone made up of a series of tight anticlines and synclines, and the Qinglonghe Fracture Zone, which is a complex NE–SW-striking, NW-dipping zone of thrust and strike-slip faults (see Fig. 2.5).

3.5.2.1 Folds and Foliation

The folds of the Qinglonghe Anticlinorium are tight and overturned to the east. The axial planes strike NE–SW and dip NW. Both the bedding and the foliation in the metamorphic rocks strike 15°–30° and dip NW at an angle of 40–60°. The Qinglonghe folds belong to the late-stage series of Sun et al. (1989), and no evidence of earlier folding in the Proterozoic rocks has been reported.

3.5.2.2 Faults and Fractures

The Qinglonghe Fracture Zone is an important structural unit affecting the entire Banbishan district. It is a NE–SW-trending (strike 5°–35°) zone of reverse and strike-slip faults. Most faults within the district strike NNE–SSW to NE–SW and dip to the NW. There are two main generations of faults judging from the mineralogy and texture of the fault zones. The older group of faults shows ductile deformation features and consists of biotite-chlorite-sericite schists and mylonites. These may be Proterozoic in age. The later faults are characterized by brittle deformation, with zones of fault breccia, cataclasite, and fault gouge. The late faults cut igneous dikes attributed to the Yanshanian Orogeny and are therefore thought to be Mesozoic in age or younger.

The largest of the early group of faults at Banbishan is a shear zone consisting of sericite-quartz schists which strikes 40°–50° and dips 50°–70° NW. The zone is 5–20 m wide and cuts through the entire mining district (Fig. 3.21). Tourmalinization is extensive in altered portions of the shear zone, and minor arsenopyrite and pyrite also occur. Another example of early faults in the Banbishan area is a chloritized mylonitic shear zone in the metaconglomerate unit of the Zhangjiagou Formation which strikes 10°–25°, dips 40°–60° NW, and is 1200 m long and 4–8 m wide. The zone contains minor gold mineralization associated with chlorite and pyrite alteration.

The most important of the late faults are designated F1, F2, and F3 (Fig. 3.21). All three are reverse faults dipping 20°–80° NW. They are characterized by meter-wide zones of fault breccia. None of the faults is significantly mineralized and, at least in the case of the F1 fault, the last movement clearly postdates mineralization and displaces a postmineralization diorite dike.

3.5.3 Gold Mineralization

3.5.3.1 Form of the Ore Bodies

The gold mineralization in the Banbishan district occurs in groups of quartz veins and alteration selvages in sheared and altered host rocks. The vein groups are associated with NE–SW-striking, NW-dipping faults. The main ore-bearing fault zone strikes 20°–40° and dips 28°–40° NW. The zone is 2–5 m wide and 1120 m long. Gold ore occurs in discontinuous quartz veins within altered and sheared wall rocks of the Shangbaichengzi Formation, and to a lesser extent in granitic dikes. The veins are 100–300 m in length and 1–2 m wide. The gold tenor ranges from 4 to 15 g/t.

3.5.3.2 Macroscopic Description of the Ores

The ores in the Banbishan deposit include both oxidized and primary types. The oxidized zone is 5–10 m thick. The ore is limonitic with a spongy texture. Apart from quartz, primary minerals are completely replaced. This type of oxidized ore has been mined out.

The primary ore can be divided into three parageneses according to the ore mineral assemblage and the nature of wall rock alteration. Most of the gold occurs in the first two types of ore.

The main ore type of the Banbishan deposit is referred to as the chlorite-silica type, and it consists of highly fractured rocks disseminated with abundant, evenly distributed quartz stringers and quartz replacements. The chlorite-silica type ore contains a relatively high-temperature mineral paragenesis with pyrite, pyrrhotite, arsenopyrite, and occasional scheelite and wolframite.

A second type of ore is termed biotite leptite type. This type is gradational between the chlorite-silica type described above and unaltered host rock. It may be very difficult to distinguish from unaltered host rock by macroscopic appearance. The ores are compact and hard due to silicification and/or contact metamorphism from igneous dikes. The ore paragenesis is the same as given above, but the concentrations are lower.

The third type of ore is termed silicified type. This type of ore has a gray-white color, contains open fractures, and is characterized by a lower-temperature ore paragenesis including pyrite, chalcopyrite, sphalerite, minor arsenopyrite, and local antimonite. Quartz occurs as massive replacements in the host rock, and as veinlets or as individual crystals in fractures. This type of ore mainly occurs in tensional parts of brittle fault zones and it is clearly Mesozoic in age or younger since the host faults cut Yanshanian igneous dikes.

3.5.3.3 Ore Petrography and Paragenesis

The primary ore minerals found in the Banbishan deposit are the following: antimonite, arsenopyrite, chalcopyrite, galena, native gold, pyrite, pyrrhotite, scheelite, sphalerite, and wolframite. No detailed description of ore petrography is available.

Gold occurs mainly in cracks within gangue (quartz, rarely feldspars) and occasionally also in cracks within pyrite and arsenopyrite. Most gold forms in the chlorite-silica and biotite-leptite type ores associated with pyrite and arsenopyrite, but it is also found associated with antimonite in the silicified-type ores. Rarely, grains are visible to the naked eye, but grain size analysis showed that about half of the native gold has a diameter of 0.5–0.04 mm and 30% has a grain size of less than 0.04 mm. The silver concentration in the gold is about 10 wt%.

3.5.3.4 Chemical Composition

The chemical composition of the ores at Banbishan is complicated by the occurrence of different types of primary and oxidized ores. The element association of the primary ores includes As, Ag, Au, Cu, Pb, Zn $\pm$W $\pm$Sb. Table 3.9 lists the range of element concentrations of 20 grab samples representing a mixture of the primary ore types.

3.5.4 Wall Rock Alteration

Three stages of wall rock alteration have been distinguished at the Banbishan deposit. From early to late these are:

a) microcline-quartz-arsenopyrite-pyrite stage,
b) quartz-pyrite-sericite-chlorite stage,
c) carbonate-pyrite stage.

The earliest stage involves the replacement of plagioclase by quartz and microcline, accompanied by early disseminated pyrite and arsenopyrite. Tourmalinization is also locally present, associated with arsenopyrite. The second stage of alteration intensifies and overprints the first. It is confined

116

Table 3.9. Concentration of selected elements in primary ore of the Banbishan deposit

Element	Concentration (ppm)	Element	Concentration (ppm)
Ag	1–30	Ni	10–30
As	1000–3000	P	1000
Au	10	Pb	15–350
Ba	200–800	Sb	100–300
Bi	3	Sr	100–700
Co	10–20	Ti	40–250
Cr	10–150	V	30–150
Cu	10–250	W	40
Mo	4–50	Zn	30–120
Mn	200–2500		

Source: unpublished analyses by Ministry of Metallurgical Industry, Beijing.

to the extensional zones within the mineralized structures. The alteration involves mainly sericitization, silicification, chloritization, and pyritization. Sericite replaces plagioclase, K-feldspar, and some biotite. The silicification is expressed by quartz replacements of all primary minerals and by the formation of quartz veinlets. Pyrite occurs both as disseminated grains in the rock and in veinlets. Chloritization occurs in the wall rock selvages as replacements of biotite and hornblende. Gold mineralization is best developed in rocks showing the sequential development of both the first and second alteration stages.

The late-stage alteration involves veining by carbonate and local pyritization along open fractures and joint surfaces. This carbonate-pyrite alteration postdates the primary gold mineralization.

3.5.5 Age of Mineralization

No isotopic dates are available to constrain the age of mineralization in the Banbishan district, but the field relations indicate the possibility of two widely separated stages of mineralization (considering only the primary, unoxidized ore). An early, pre- or synmetamorphic stage is suggested by the fact that weak mineralization is present in ductile, greenschist-facies mylonite schist zones. It is possible, however, that the mineralization postdates the formation of the mylonites, and detailed petrographic studies are needed to determine if this is the case. In any case, most of the mineralization, and all of the economic ore at Banbishan, consists of vein stringers and disseminated zones in brittle structures which clearly cut undeformed igneous dikes and are therefore postmetamorphic. The vuggy nature and low-temperature ore assemblages of the silicified type of ore indicate that the late stage of mineralization continued under near-surface conditions.

4 Aspects of Metallogenesis

The similarities in age, geologic setting, and ore association in the various gold districts of eastern Hebei province documented in Chapter 3 suggest the likelihood of a common origin. The purpose of this chapter is to synthesize data and observations from all of the deposits studied in an attempt to constrain their origin. The chapter is organized according to the six metallogenetic questions posed in the introduction, namely:

4.1 The age of mineralization.
4.2 The source of gold.
4.3 The role of iron-rich host rocks.
4.4 The role of granites.
4.5 Structural controls.
4.6 The nature of ore-bearing fluids and conditions of mineralization.

The following discussions are based partly on information from Chapters 2 and 3, and partly on data not previously presented in the text.

4.1 The Age of Mineralization

4.1.1 Field Relations

Field evidence concerning the age of mineralization consists of the nature of the ore-bearing structures (geometry and type of deformation) and cross-cutting relationships between ore veins and igneous intrusions (dikes, plutons) of known or inferred age. The field relations in all of the deposits studied suggest that the main stage of mineralization postdated regional metamorphism. The main-stage ore veins are discordant to the host rock structures and show brittle deformation features. However, it must be noted that there is evidence for a synmetamorphic stage of mineralization in the Jinchangyu and the Banbishan gold deposits.

In the Jinchangyu deposit, the early stage of mineralization is represented by folded and concordant quartz-albite-sulfide veins, and by mylonitic fabrics and greenschist-facies mineral assemblages in some mineralized zones. L.S. Yang (1988) attributes the synmetamorphic mineralization to the Early Proterozoic Fuping Orogeny (ca. 2500 Ma). This early stage of mineralization at Jinchangyu is subeconomic. The main stage of mineralization, which formed most of the ore bodies and/or upgraded the earlier-formed veins, is represented by quartz veins and altered fault zones which are discordant to the host rock foliation and locally cross-cut undeformed igneous dikes of Yanshanian age (L.S. Yang 1988). In the large shear zone-hosted ore bodies at Jinchangyu, both early and late stages of mineralization may be superimposed.

The gold deposit at Banbishan also shows evidence of two mineralization periods. An early stage is suggested by weakly mineralized, ductile, mylonitic shear zones with greenschist facies mineral assemblages. Detailed petrographic studies of these zones are lacking, however, and it is presently unknown if the mineralization is contemporaneous with the deformation and metamorphism or if it formed later. In any case, the main mineralization at Banbishan, which reaches economic grade, consists of quartz stringers and disseminated zones in semi-brittle faults which crosscut undeformed igneous dikes attributed to the Yanshanian Orogeny. This main stage of mineralization is therefore Mesozoic in age.

In the Yuerya, Niuxinshan, and Sanjia districts there is no evidence for synmetamorphic mineralization. The mineralized veins cross-cut Yanshanian granite intrusions and/or dikes, some of which have been radiometrically dated at 140–170 Ma. Therefore the maximum age of mineralization in these cases is Jurassic. As discussed later in Chapter 4.5.3, the orientation of mineralized structures in these deposits also suggests that mineralization occurred as a result of the Yanshanian Orogeny.

4.1.2 Isotopic Age Dates

The available isotopic ages of hydrothermal vein minerals from the gold districts of eastern Hebei province are summarized in Fig. 4.1 along with all available ages from Yanshanian granites and dikes in the region. Jurassic ages from hydrothermal minerals have been found in all districts for which data are available. It is remarkable that all of the mineral ages from the Niuxinshan, Wangtoushan, Huajian, and Jinchangyu deposits indicate a mineralization event between 170 and 190 Ma. Based on the few cases where ages from hydrothermal minerals and granites exist from the same district, the hydrothermal minerals are the same age as, or slightly younger than the granites.

The age resolution of the data in Fig. 4.1 is rather poor, but the data suffice to demonstrate that gold mineralization took place during the Yanshanian Orogeny near the time of emplacement of granitic magmas. Further dating studies should attempt to establish a precise chronology of magmatism and gold mineralization. An important step would be to date the igneous dikes which were emplaced after the granites but before the formation of ore veins.

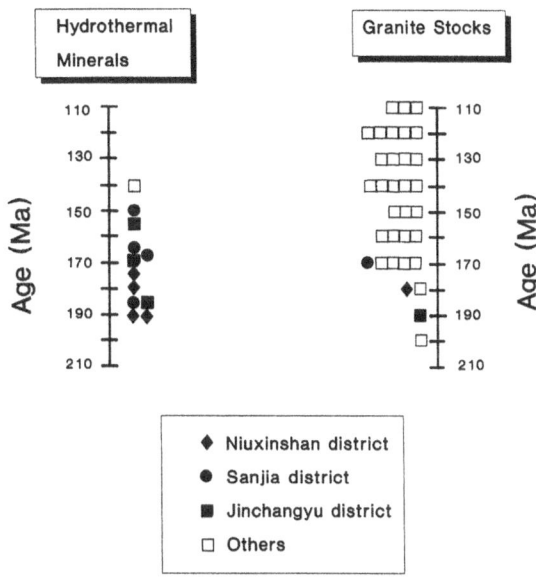

Fig. 4.1. Summary of isotopic age data from hydrothermal minerals and Yanshanian granites in the gold districts of eastern Hebei province described in the text

4.2 The Source of Gold

The source of metals in a given deposit or mining district is one of the most elusive questions in metallogenesis but also one of the most important because of its obvious strategic significance in exploration. The geologic characteristics and age of the gold deposits of eastern Hebei province (Chap. 3) constrain the possible sources of gold practically to three:

1. The gold was leached from the Precambrian metamorphic rocks which surround and/or underlie the deposits.
2. The gold was derived from the Yanshanian granites.
3. The gold was derived from the mantle, and it was transported by juvenile fluids along crustal-scale fault zones.

4.2.1 Gold in the Qianxi Group: The "Source Bed" Concept

One of the most striking features of the gold deposits discussed in this text, and many of the deposits in northeastern China generally, is that they are located in mafic to intermediate high-grade metamorphic rocks of Archean to Early Proterozoic age (Fig. 1.3). Zhu (1989) has estimated that over 50% of all gold deposits in the whole of China are hosted by Early Precambrian metamorphic rocks. The close association of gold with Early Precambrian

121

host rocks in northeastern China suggests that these rocks may be the source of gold. The "source-bed" concept proposes that particular units or beds within the host rocks are gold-rich and, if identified, may guide exploration. There are many proponents of the "source-bed" concept in relation to the metallogenesis of gold in northeastern China. Zhu (1985) pointed out that, in the case of the complex Precambrian basement of northeastern China, "primary" source beds and "derived" source beds should be distinguished, the latter being formed by metamorphic or sedimentary reworking of the former. Guan (1988) surveyed the data on gold concentrations in the Precambrian basement of Liaoning province and concluded that the Archean rocks are depleted in gold relative to similar rocks in greenstone belts of South Africa and Canada, whereas the Early Proterozoic rocks are considerably enriched in gold relative to the crustal average. He suggested that the Early Proterozoic rocks are "secondary source beds" which reworked and concentrated the gold which was presumably present in their Archean predecessors. An interesting further variation on the source-bed concept was discussed by Yang (1989) based on insights provided by the deep drilling project of the Kola Peninsula. Yang pointed out that fluid-saturated zones of high fracture density occur in the Kola well from the 4.9-km level down to about 9 km. The zones are capped by relatively impervious rocks, and they therefore pond crustal fluids. Yang (1989) termed such structurally controlled zones of metal-rich hot brines "deep liquid ore source".

In past efforts to search for a gold-enriched source bed in the Precambrian basement of northeastern China, it proved to be very difficult to reliably determine representative "background" gold concentrations of the various rock units. The reason is that the rocks underwent several metamorphic and tectonic events. Even when interesting gold values are analyzed in apparently unmineralized samples, it is difficult to demonstrate whether these concentrations are the "primary" gold values inherent to the protolith or are due to enrichment or depletion from later metamorphism or metasomatism. This problem can be illustrated from the previously published analyses of "background" gold concentrations in the Qianxi Group rocks from the Jinchangyu district. Zhu (1985) reported an average gold concentration of 71 ppb (21 samples analyzed), which is a tenfold enrichment over the average of 3 to 7 ppb for crustal igneous and metamorphic rocks (Boyle 1979; Crocket 1991). On the other hand, Gao and Lin (1987) reported an average gold concentration of 3.1 ppb (14 samples analyzed) from rocks in the Jinchangyu district. The fundamental question is how to interpret these results. Are the high values representative for the gold concentration in the protolith or have they been influenced by secondary processes?

We consider it very likely that the high Au concentrations reported from the Qianxi Group rocks reflect secondary gold enrichment, which is difficult to avoid if samples are taken in the mining districts. One indication of this is that recent publications consistently report low values for gold concentrations in the Archean rocks collected outside of the mining districts. Yu and

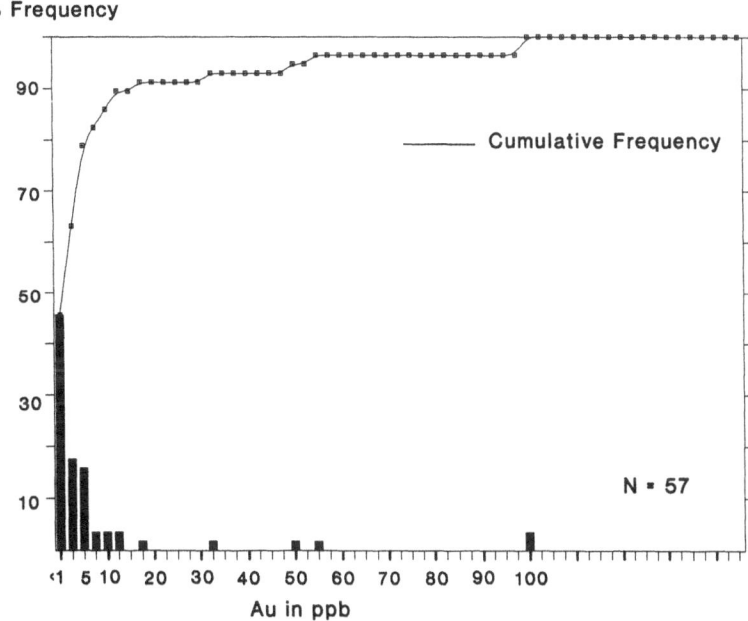

Fig. 4.2. Histogram and cumulative frequency curve of gold concentration in samples of the Archean Qianxi Group collected from outside of the mining districts in eastern Hebei province

Jia (1989) report gold concentrations ranging from 0.5 to 2.3 ppb from 65 samples representing several lithologic units in the Upper Qianxi Group. Yu et al. (1989) reported an average gold concentration of 5 ppb from 55 samples of the Qianxi Group rocks near the Jinchangyu district.

Our own analytical results from 57 samples of the Qianxi Group rocks are presented in Appendix 2. The samples were taken along an east-west profile through the Qianxi Group, located well to the south of the mining districts near Zunhua (see Fig. 1.3). The samples represent a range of rock types from pyroxenites to anorthosites and magnetite quartzites. The gold concentrations are shown in a cumulative frequency diagram in Fig. 4.2. The figure shows that over 80% of the samples have gold values at or under 5 ppb and the rest scatter to higher concentrations, up to a maximum of 100 ppb. A similar distribution of host-rock gold concentrations has been reported, for example, by Saager and Meyer (1984) from the Barberton area of South Africa. These authors suggested that the low "background" gold concentration (geometric mean 1.3 ppb) represents gold contained in the rock-forming minerals, and that the few abnormally high gold values reflect the presence of gold-rich sulfides or native gold on grain boundaries or microfractures. This would also be a reasonable explanation for the distribution of gold in the Qianxi Group rocks.

Primary Gold Concentrations and the Au-Pd Method

The work of Keays (1984) showed that the gold concentrations measured in metamorphic rocks in greenstone belts are generally not representative of their primary values. Keays and Scott (1976) demonstrated the problem from a study of gold in ocean ridge basalts. They found higher gold concentrations in the glassy margins of basalt pillows than within the spilitized pillow interiors. The gold concentration in the glassy margins was considered to represent that of the quenched lava, whereas the lower concentration in the spilitized pillow interiors was attributed to leaching by the seafloor metamorphism. The implications of this to the problem of determining primary gold concentrations in the high-grade metabasites of the Qianxi Group are disheartening, since these rocks have undergone not only probable seafloor metamorphism, but also at least two periods of high-grade regional metamorphism.

Keays (1984) suggested a potential method to recalculate the primary gold concentrations in altered or metamorphosed rocks based on the geochemical behavior of Au and Pd. He argued that Au and Pd occur in a nearly constant ratio in mafic and ultramafic magmas whereas Au is much more mobile than Pd during metamorphic and metasomatic processes. Keays suggested that the primary Au/Pd ratio of komatiitic magma is 0.37 based on analyses of glassy ultramafic lavas (picrites) of Gorgona Island, Colombia and from Disko Island, Greenland. He then showed that the measured Au/Pd ratios of Archean komatiites from Canada (Munro Township), Zimbabwe (Belingwe), Western Australia (Kambalda and Mt. Clifford), and South Africa (Barberton) are generally lower than the "primary" value, suggesting preferential mobilization of gold. The degree of gold "loss" correlates roughly with the degree of alteration and metamorphic grade of the rocks. Wang (1988) introduced the "gold-palladium method" into the Chinese literature. He used Keays' (1984) results from Archean komatiites in Western Australia and South Africa as evidence that up to 75% of the primary gold in such rocks can be leached after crystallization of the protolith. Wang concluded that the primary gold contents of metamorphic rocks should be recalculated before discussing the source of gold.

Although the qualitative value of the Au-Pd method in distinguishing primary and disturbed noble metal signatures is established, the method may not give a reliable quantitative estimate of primary gold concentrations. One important source of uncertainty is the assumed initial Au/Pd ratio. The primary ratio of 0.37 given by Keays (1984) was based on an average value from analyses of natural picrites which, in fact, showed considerable variation in Au and Pd concentrations. Another source of error is the assumption that Pd is immobile during metamorphism. This seems to be

124

Table 4.1. Gold and palladium concentrations in ultramafic samples from the Qianxi Group, and their recalculated "primary" gold concentrations

Sample	Rock type	SiO$_2$	TiO$_2$	Al$_2$O$_3$	Fe$_2$O$_3$	MgO	MnO	CaO	Na$_2$O	K$_2$O	P$_2$O$_5$	Au	Pd	Au/Pd	"Primary Au"
CE102	Websterite	43.88	0.17	5.43	11.44	27.34	0.11	4.63	0.64	0.36	0.18	0.5	32.0	0.02	11.84
CE098	Pyroxene amphibolite	42.98	0.22	5.06	11.58	26.86	0.07	5.88	0.30	0.08	0.05	0.5	28.0	0.02	10.36
CE097	Websterite	42.06	0.23	6.88	15.32	25.52	0.12	5.38	0.34	0.18	0.09	6.0	45.0	0.13	16.65
CE099	Websterite	52.64	0.14	3.62	9.43	24.84	0.12	8.01	0.30	0.08	0.05	10.0	67.0	0.15	24.79
CE101	Websterite	53.18	0.19	3.59	9.90	24.56	0.10	6.76	0.54	0.18	0.01	0.5	14.0	0.04	5.18
YQ8	Pyroxenite	–	0.82	–	16.20	21.56	0.15	7.00	1.05	0.30	0.14	50.0	2.0	25.00	0.74
CE100	Pyroxenite	51.62	0.19	4.82	9.70	18.93	0.10	10.38	1.28	0.28	0.01	0.5	18.0	0.03	6.66
CE201[b]	Hornblendite	49.80	1.46	6.29	15.32	17.96	0.14	6.70	1.32	0.12	0.04	32.0	2.0	16.00	0.74
CE208	Pyroxenite	49.64	0.42	4.05	11.88	17.34	0.18	12.57	0.99	0.25	0.02	0.5	10.0	0.05	3.70
CE209[a]	Hornblendite	47.98	0.50	7.81	12.63	17.10	0.15	11.74	1.27	0.50	0.09	0.5	2.0	0.25	0.74
CE106[a]	Hornblendite	43.78	1.99	6.20	18.05	16.83	0.12	7.59	0.80	0.28	0.23	0.5	5.0	0.10	1.85
CE093[a]	Pyroxene amphibolite	43.06	1.85	5.26	19.80	16.67	0.13	8.75	1.22	0.28	0.46	0.5	3.0	0.17	1.11
CE103	Pyroxene amphibolite	43.96	1.48	6.09	17.36	16.39	0.18	9.25	1.12	0.08	0.16	0.5	4.0	0.13	1.48
CE095[b]	Hornblendite	44.12	1.89	4.68	19.05	15.40	0.12	8.63	1.04	0.18	0.25	98.0	2.0	49.00	0.74
CE096[a]	Hornblendite	44.18	1.77	6.10	17.64	15.40	0.10	9.75	0.88	0.64	0.27	1.0	2.0	0.50	0.74
CE104[a]	Pyroxene amphibolite	41.56	2.17	6.85	19.04	15.29	0.34	10.75	1.04	0.46	0.34	2.0	5.0	0.40	1.85
Q25	Pryoxene hornblendite	48.40	0.65	9.40	9.80	14.65	0.15	12.93	1.22	1.08	0.21	2.0	11.0	0.18	4.07
CE202	Pyroxenite	52.12	0.78	8.44	9.80	13.17	0.12	11.99	2.04	1.40	0.09	2.0	2.0	1.00	0.74
CE105[a]	Hornblendite	43.75	2.02	7.59	17.93	12.97	0.46	12.12	1.46	0.64	0.46	2.0	5.0	0.40	1.85
YQ108	Hornblendite	–	0.73	–	15.70	12.10	0.15	9.79	2.58	0.90	0.48	1.0	2.0	0.50	0.74
YQ124	Amphibolite	–	0.43	–	13.70	11.61	0.13	10.91	1.77	0.42	0.05	1.0	10.0	0.10	3.70
YQ88	Pyroxenite	–	0.62	–	14.70	11.11	0.17	15.39	1.73	1.02	1.81	98.0	3.0	32.67	1.11
YQ123	Pyroxenite	–	0.47	–	12.40	10.94	0.13	7.28	2.78	0.45	0.11	1.0	2.0	0.50	0.74
YQ41	Pyroxenite	–	0.37	–	12.30	10.45	0.13	11.19	2.75	1.01	0.05	2.0	3.0	0.67	1.11
YQ45	Pyroxene amphibolite	–	0.40	–	12.60	10.11	0.12	10.49	2.87	1.06	0.05	3.0	2.0	1.50	0.74

Analyses by XRF and INAA following fire assay (Au) or nickel sulfide (Pd) preconcentration, total Fe as Fe$_2$O$_3$.
"Primary Au" = Au × (Au/Pd × 0.37), see text for details.
[a] In contact with BIF lens.
[b] Sample near a fracture zone.

justified for sea-floor metamorphism according to studies of Keays and Scott (1976) and more recent work on precious metals in oceanic basalts by Brugman et al. (1987) and Hamlyn et al. (1985), but the situation may be very different in the high-grade polymetamorphic conditions to which the Archean rocks of northeastern China have been subjected.

We have applied the Au-Pd method to better interpret the Au concentrations of the Qianxi Group rocks. Only samples with MgO contents over 10 wt% were used, because in most other samples, the Pd concentrations are at or below the detection limit of 1 ppb, and because the bulk composition of these rocks justify the application of the "primary" Au/Pd ratio (0.37) derived from picritic basalts and komatiites. Table 4.1 presents the results from 25 samples of the Qianxi Group. Listed in the table are the bulk composition, measured Au and Pd values, Au/Pd ratios, and the calculated "primary" Au concentrations.

The following points can be concluded from Table 4.1:

a) Both the Pd and Au concentrations measured in these samples are highly variable and the Au/Pd ratio is not constant. If the "primary" Au/Pd ratio of 0.37 (or any other fixed ratio) is appropriate for the protoliths of these rocks, then postcrystallization mobility of Au and/or Pd can be inferred.

b) In most samples, the calculated "primary" gold concentrations are higher than the measured concentrations, which suggests leaching of gold (subject to the assumptions given above).

c) The highest measured Pd concentrations are in samples of ultramafic pyroxenites. The gold concentrations in these samples are not correspondingly high, suggesting possible leaching of primary gold from the rocks after crystallization.

d) The highest measured gold concentrations occur in samples taken near fractures. Pd is not correspondingly high in these samples, suggesting possible gold enrichment by postcrystallization processes acting along rock fractures.

e) Most importantly, neither the measured nor the calculated "primary" gold concentrations are significantly higher than the crustal average.

4.2.2 Gold in Yanshanian Granites

The most compelling arguments for the Yanshanian granites (sensu lato) as the source of gold are, first, that the mineralization in most mining districts studied is broadly coeval with the Yanshanian magmatism; and second, the gold deposits are spatially associated with Yanshanian granite plutons and related dikes.

There are much fewer data on gold concentrations from the Yanshanian granites in eastern Hebei province than from the Qianxi Group rocks. Table 4.2 lists the available data from several Yanshanian intrusions based on

Table 4.2. Gold concentration in unaltered samples of granitic plutons associated with mineralization in eastern Hebei province

Name of intrusion	Number of samples	Average Au (ppb)	Reference[a]
Niuxinshan	3	5.3	1
Wangtoushan	1	3	1
Sanyihe	4	3	1
Yuerya (white)	4	4.4	2
Yuerya (red)	4	1.9	2
Maoshan	32	2.5	3

[a] 1, This study, analyses by XRAL laboratories, Ontario, by fire assay, and DCP; 2, Yu et al. (1989), analytical method not given; 3, Yu and Jia (1989), analytical method not given.

our own analyses and on published studies. All of the plutons listed are directly associated with gold deposits, and some are mineralized themselves (Niuxinshan, Yuerya).

The data in Table 4.2 represent only a small number of samples from a few plutons, and more data need to be collected before a definitive statement about the gold potential of the Yanshanian granites can be made. However, at present it appears that the gold concentrations in the Yanshanian granites are less than 5 ppb, i.e., gold does not appear to be enriched in the granites relative to either the crustal average (Boyle 1979; Crocket 1991) or to the Qianxi Group rocks (Fig. 4.2, Appendix 2).

The interpretation of these data can be questioned because the rock samples analyzed do not represent the original magma from which they crystallized, especially because of the loss of volatiles. As discussed by Burnham and Ohmoto (1980), for example, much of the Cl and S in a magma can be removed at a late stage of crystallization by exsolution of volatiles. The importance of Cl and S as ligands for gold complexing suggests that the gold concentration may depend strongly on the behavior of these volatiles in the magma.

4.2.3 Isotopic Constraints

The source of gold may be indirectly constrained by considering isotopic evidence for the source of the ore elements lead and sulfur which accompany gold in the deposits. The following discussion is based on lead and sulfur isotopic analyses of ore and gangue minerals from selected deposits in eastern Hebei province.

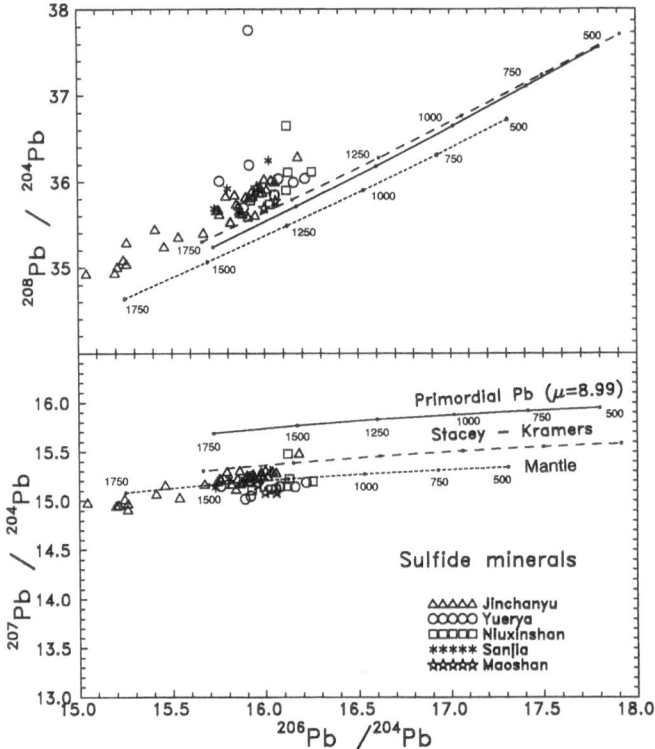

Fig. 4.3. Lead isotopic composition of galena and pyrite from selected gold deposits in eastern Hebei province. *Numbers* indicate age in Ma. Data sources are given in the text

Common Lead Isotopes

The use of common lead isotopes to deduce the source of lead in ore deposits is well established (Doe and Stacey 1974; Doe and Zartman 1979; Kramers and Foster 1984). Previous work on lead isotopes from ore deposits in China is reviewed by Zhu and Chen (1984) and Wang (1989). Wang (1989) concluded that the ore lead in the metamorphic-hosted gold deposits in northeastern China "is mostly derived from old multistage lead of a lower crustal uranium-depleted source." This statement also seems to hold true for the gold deposits from eastern Hebei province discussed below.

Figure 4.3 shows a compilation of lead isotopic data from ore minerals (mostly galena, also pyrite) from the gold deposits of Jinchangyu, Yuerya, Niuxinshan, and Sanjia. Figure 4.4 presents the corresponding lead isotopic compositions of K-feldspars and whole-rock samples from the Yanshanian granites in the respective mining districts. The diagrams combine data from Lin (1985), Yu and Jia (1989) and Yu et al. (1989), together with

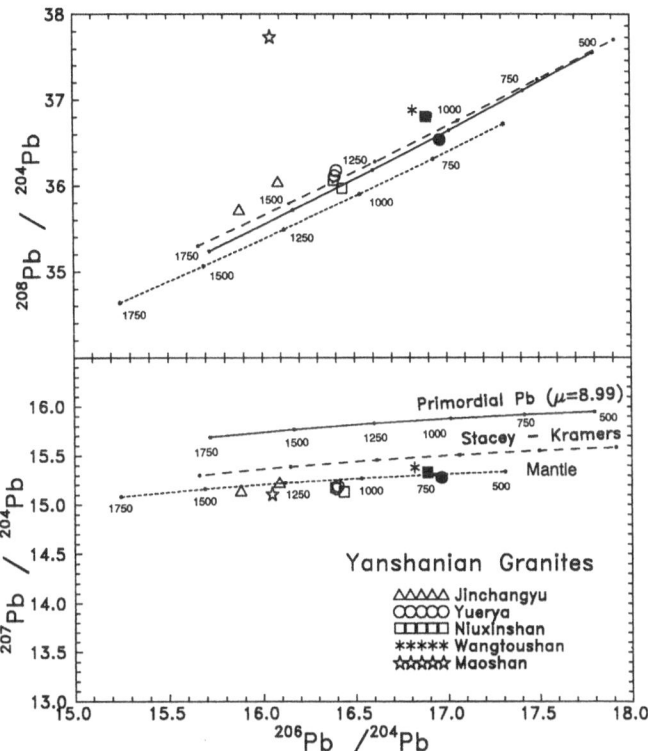

Fig. 4.4. Lead isotopic composition of Yanshanian granite K-feldspar (*open symbols*) and whole rocks (*filled symbols*) from selected gold deposits in eastern Hebei province. *Numbers* indicate age in Ma. Data sources are given in the text

unpublished analyses of the Ministry of Metallurgical Industry. Also shown in each diagram are single-stage lead isotopic growth curves for primeval lead (after Yu and Jia 1989), mantle lead (after Doe and Zartman 1979), and the two-stage terrestrial lead growth curve of Stacey and Kramers (1975).

The ore from all of the deposits shown has quite similar lead isotopic compositions. Except for some samples from the Jinchangyu deposit, the values of $^{206}Pb/^{204}Pb$, $^{207}Pb/^{204}Pb$ and $^{208}Pb/^{204}Pb$ for the ore leads cluster around 16, 15, and 36, respectively. The lead data from the Jinchangyu deposit overlap this range, and extend to lower, less-radiogenic values of $^{206}Pb/^{204}Pb$ and $^{208}Pb/^{204}Pb$. None of the deposits contains conformable, single-stage lead, and all of the deposits contain very nonradiogenic lead for their age; the lead isotopes yield geologically meaningless single-stage model ages (based on Doe and Stacey 1974) between about 1200 and 2000 Ma, whereas the deposits themselves are Mesozoic (see Chap. 4.1). These lead isotopic compositions resemble the examples discussed by Doe and Zartman

(1979) from ore deposits in the "rejuvenated craton" environment. Such lead tends to be low in the uranogenic isotopes ^{207}Pb and ^{206}Pb, and rich in the thorogenic isotope ^{208}Pb compared with crustal, mantle, or average terrestrial lead. Doe and Zartman (1979) suggest that this type of lead could be derived from lower crustal rocks which were depleted in uranium due to granulite-facies metamorphism. This interpretation seems reasonable for the Archean granulite-facies metamorphic terrane of eastern Hebei Province.

Yu and Jia (1989) interpreted the isotopic composition of ore leads in eastern Hebei province based on a two-stage model. The lead isotope data from the Jinchangyu and Yuerya deposits form linear arrays on the diagram of ^{207}Pb/^{204}Pb against ^{206}Pb/^{204}Pb, which the authors interpreted as secondary isochrons. The intersections of the secondary isochron from the Jinchangyu deposit with the primeval lead growth curve indicate model ages of 3480 Ma for the lead source and 133 Ma for the mineralization. The corresponding model ages from the secondary isochron from the Yuerya deposit are 3850 Ma and 230 Ma. This interpretation is in general agreement with the geologic facts (i.e., Early Archean basement, Mesozoic mineralization) but the ages are highly model-dependent and should be considered qualitative at best.

The isotopic composition of lead from the Yanshanian granites associated with the gold deposits also gives important evidence bearing on the source of metals. The comparison of Figs. 4.3 and 4.4 shows that the isotopic compositions of lead from the ores and from the granite K-feldspars are very similar, whereas the lead from granite whole-rock samples is slightly more radiogenic. The more radiogenic whole-rock lead is expected because it can continue to evolve by decay of U and Th after crystallization of the rock, whereas the common lead in K-feldspar is isolated from U and Th. The important point to be made is that the lead from the granites, like that from the ore deposits, is nonconformable, anomalous, and nonradiogenic for its (Mesozoic) age. This suggests that the lead in the granites is also derived, at least in large part, from uranium-depleted metamorphic basement rocks.

In summary, the lead isotopic data from both the granites and from the ores rule out an important contribution of "normal" Mesozoic lead (in the sense of Doe and Stacey 1974) in the deposits studied. The ultimate source of lead must be the uranium-depleted metamorphic basement. Of course, the source of lead in these deposits need not be the same as the source of gold, but there is a strong paragenetic association between gold and galena in the ores.

Sulfur Isotopes

The isotopic composition of sulfur in sulfide minerals can, in favorable cases, constrain the source of sulfur in the ores. Figure 4.5 summarizes the sulfur isotopic compositions of pyrite from five gold deposits in eastern Hebei province taken from Yu and Jia (1989) and Yu et al. (1989). Also shown in the diagram (open circles) are whole-rock δ^{34}S values from

130

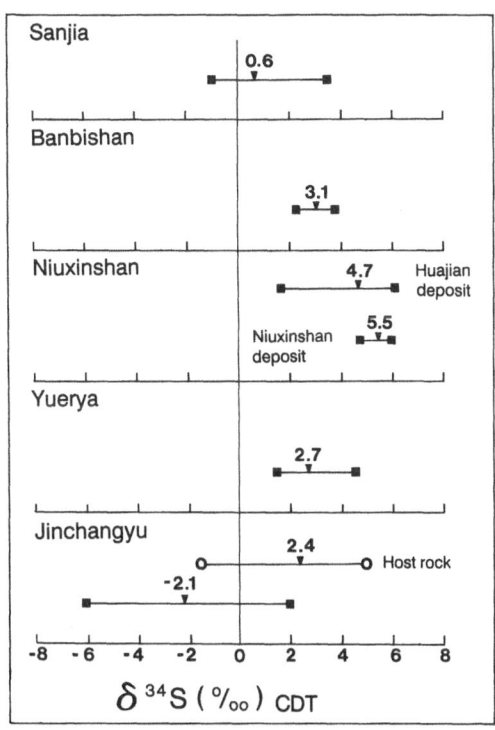

Fig. 4.5. Sulfur isotopic composition of pyrite (*filled squares*) and host rock (*open circles*) from selected gold deposits in eastern Hebei province. *Numbers* indicate mean values. Data replotted from Yu and Jia (1989) and Sun et al. (1989)

amphibolite country rocks from the Jinchangyu district. The pyrites from all of the deposits and the whole-rock amphibolite sample contain sulfur with a narrow range of $\delta^{34}S$ values between -6 and $+6‰$.

The isotopic compositions shown in Fig. 4.5 are typical for sulfide compositions in mesothermal gold deposits from many parts of the world (Lambert et al. 1984; Kerrich 1987; Peters and Golding 1989; Groves and Foster 1991). Yu and Jia (1989) argued that the similar, low $\delta^{34}S$ values of ore sulfides and amphibolites from the deposits in eastern Hebei province indicate that the sulfur was derived from the Archean basement. This interpretation is geologically reasonable and consistent with the lead isotopic data discussed above, but it must be emphasized that the sulfur data do not rule out other possibilities. According to Ohmoto and Rye (1979) and Ohmoto (1986), such $\delta^{34}S$ values are consistent with a deep-seated source of sulfur either from leaching of metamorphic rocks or from granitoid intrusions. In most rocks, such low values for $\delta^{34}S$ would rule out oxidized oceanic sulfur as a source (modern ocean sulfate has $\delta^{34}S$ values of $+20‰$). However, as Lambert et al. (1984) pointed out, the situation is different for Archean rocks because the sulfur isotopic composition in the reduced Archean hydrosphere would have had $\delta^{34}S$ values near 0. In addition, the isotopic composition of sulfide minerals depends in large part on the solution chemistry from which they crystallized, in particular the pH and oxygen fugacity, and these parameters are poorly defined for the ores at present. Therefore, the sulfur isotopic data from the ores must be considered ambiguous as to the sulfur source. If one makes the reasonable assumption that the source of sulfur, whatever it was, had $\delta^{34}S$ near zero, then the low positive $\delta^{34}S$ values of pyrite from most of the deposits suggest that the mineralizing fluids were chemically reduced.

Comparing the data from the different eastern Hebei deposits in Fig. 4.5, the Jinchangyu deposit is exceptional in having pyrite with mostly negative values of $\delta^{34}S$ whereas all others show positive values. Assuming that the sulfur source for all of the amphibolite-hosted deposits studied was the same, then the unusually negative $\delta^{34}S$ values of pyrite from the Jinchangyu deposit can be explained simply by a higher oxidation state of the mineralizing fluids in that deposit. Because sulfate concentrates ^{34}S relative to sulfide (Ohmoto 1986), the pyrite precipitating from an oxidized fluid will be isotopically lighter than that from a more reduced fluid, other conditions being equal.

4.2.4 The Mantle Connection

The significance of mantle processes for metallogenesis is a topic of current debate in studies of the Archean gold deposits in Western Australia and the Canadian Shield (Colvine 1989; Fyon et al. 1989; Groves et al. 1989; Perring et al. 1989). Most metallogenetic models which favor the mantle connection do not imply that the gold itself is derived from the mantle. Rather, the role of the mantle is simply to set the metallogenetic process in motion by the introduction of heat and volatiles to the lower crust. This causes granulitization and partial melting of a large volume of lower crustal rocks. The upward transport of metamorphic and/or juvenile fluids and deep-seated magmas has the potential to leach and transport gold. The focusing of fluid flow along regional shear zones provides a mechanism to concentrate gold in upper crustal levels and to ultimately form deposits. Such models offer an explanation for several very common features of mesothermal gold deposits, namely, the occurrence of CO_2-rich, low salinity fluid inclusions, the large-scale carbonatization of wall rocks, the proximity of deposits to regional shear zones with deep-seated intrusive rocks (lamprophyres) and granites, and the mantle signature of stable isotopic data (C-S-O).

The gold deposits in eastern Hebei province have many features in common with those of the Australian and Canadian gold deposits mentioned above, and the ubiquitous presence of lamprophyres in the gold districts described in Chapter 3 indicate the presence of deep-seated magmas closely following the time of mineralization. On the other hand, the lead isotopic evidence discussed above rules out any direct input of lead from the mantle at the time of mineralization. Mantle processes played an important role in the original formation of the Early Archean amphibolites and in their granulite-facies metamorphism around 2.5 to 2.7 Ga (Chap. 2.2.1). However, the main stage of gold mineralization in all deposits studied is Mesozoic in age, and there is no evidence in any of the deposits for Mesozoic lead of either mantle or crustal origin. A further problem is that, if mantle-derived fluids and/or magmas ascendent along major fault zones were important for gold mineralization, then it is odd that the gold in eastern Hebei province occurs almost exclusively in Early Precambrian rocks whereas the major fault zones which might tap mantle fluids are much more common in the areas of Late Proterozoic and Phanerozoic exposures (see Figs. 1.3 and 2.6).

4.2.5 Conclusions

We do not accept the concept of gold-rich source beds within the Qianxi Group. Recent studies of gold concentrations in Early Proterozoic rocks (Zhou 1989) and in the Archean Qianxi Group (this study; Yu and Jia 1989) suggest that these rocks have normal gold concentrations, and that the occasional high values encountered are due to secondary enrichment. Use of the Au-Pd method (Table 4.1) to recalculate the primary, i.e., pre-metamorphic Au concentrations in ultramafic rocks of the Qianxi Group results in values which are still not above the crustal average for such rocks.

The Yanshanian granites are spatially and temporally associated with the gold deposits in eastern Hebei province. They have been less well studied than the Qianxi Group rocks, but available data suggest that the granites, too, contain unexceptional gold concentrations (1–5 ppb) which are in the same range as those of the Archean rocks. In terms of their absolute gold abundances, then, both the granites and the Qianxi Group rocks would be equally favorable as a gold source. However, the lead isotopic data indicate that the lead in the gold deposits is derived from the Archean basement and not from the granites. Furthermore, the Archean rocks constitute a more likely source of gold than the granites simply because they are far more abundant, and they have a higher secondary permeability due to folds, faults, and fractures.

The possibility that gold was derived from the mantle during the Yanshan Orogeny cannot be entirely ruled out, but it is inconsistent with the lead isotopic evidence and it cannot account for the fact that gold deposits are concentrated in the areas of Archean exposures.

We conclude that the gold in the deposits studied was mainly derived from leaching of the Archean metamorphic basement rocks despite the fact that they had average gold concentrations less than 5 ppb. We envision this leaching as taking place in the brittle regime at the intrusion level of the Yanshanian granitoids where fluid circulation is enhanced by secondary permeability. There is abundant support in the literature for the concept that rocks with low-ppb abundances of gold can be the source of gold for economic deposits (Fyfe and Kerrich 1984; Keays 1984; Phillips et al. 1987; Groves and Foster 1991). Seward (1984), for example, reported analyses of sulfide precipitates from several active geothermal systems in New Zealand which show that ore-grade gold concentrations of 50–80 g/t can precipitate from fluids with low-ppb or even sub-ppb concentrations of gold in solution. The problem of a gold-rich source may, therefore, be less important to the genesis of the deposits than the problem of a large volume of fluid flow and a suitable chemical environment for gold leaching, transportation, and precipitation.

4.3 The Role of Iron-Rich Host Rocks

The Archean and Early Proterozoic metamorphic host rocks in many of the gold districts in northeastern China contain supracrustal sequences including metamorphic banded iron formation (BIF). The nature of these BIF rocks has been recently reviewed by Sills et al. (1987b) and Zhai and Windley (1990). The association of gold with BIF in other Archean cratons in the world has been noted by many authors (Neall 1987; Saager et al. 1987; Groves et al. 1988; Colvine 1989; Foster 1989) but has not been given much attention in northeastern China. For this reason, a discussion of Archean iron formations and their relevance to gold mineralization is appropriate. After a general introduction, the iron formations in the gold districts of eastern Hebei province are described in detail.

Two main classes of iron formations are distinguished based on examples from North America: (1) the Algoma-type, and (2) the Superior-type (Guilbert and Park 1986). The Algoma-type BIFs are directly associated with volcanic rocks, and the iron enrichment is considered to be related to submarine volcanic processes. Algoma-type iron formations are typically of Archean age, although some more modern examples are known. They tend to be smaller and less continuous than the Superior-type. The Superior-type BIFs are typically of early to Middle-Proterozoic age and, although the rock association may include some volcanic input, this is often minimal or absent. The associated rocks indicate a stable shelf depositional environment, and the Superior-type iron formations are often of vast extent. A further important criterion for classifying iron formations is the facies concept. Four facies are distinguished according to the dominant mineralogy of the iron ore, i.e., oxide, silicate, sulfide, and carbonate facies. The type of facies is related to the depositional setting of the iron formations, especially the water depth. The depth of the paleobasin increases in the order oxide-carbonate-sulfide facies (Guilbert and Park 1986).

Gold mineralization in many of the world's Archean cratons, particularly in the greenstone-granite terranes, is closely associated with Archean BIF of Algoma-type. The Superior-type BIF deposits, on the other hand, are rarely mineralized with gold to a significant extent. The metallogenesis of BIF-related gold mineralization is currently disputed (Saager et al. 1987; Foster and Gilligan 1987; Lhotka and Nesbitt 1989; Oberthür et al. 1990). Based on the observations and interpretations given in these studies, it appears that the gold mineralization associated with BIF is obviously stratabound but not always stratiform, being often localized in texturally late quartz stringers, along axial planes of folds or in structural dilatant zones. The formation of ore bodies of economic interest appears to be the product of a polystage process including syn-sedimentary gold enrichment (with barium, arsenic, sulfur), diagenetic reactions, and finally metamorphic and/or postmetamorphic hydrothermal remobilization. Perhaps the most important fact underlying this complexity is that iron-rich units can act from the very beginning of their formation as an effective geochemical trap for gold in both the sedimentary and hydrothermal environment.

4.3.1 Banded Iron Formations in Northeastern China

The distribution, metamorphic grade and geotectonic setting of the Archean and Early Proterozoic BIFs in China have been discussed in some detail by Sills et al. (1987b) and by Zhai and Windley (1990), and the reader is referred to those references for general information. Zhai and Windley (1990) emphasized the fact that many of the BIFs in northeastern China are found in the Archean high-grade granulite terrane and that this rock association is rare in other cratons, a notable exception being Isua in western Greenland. In the Archean rocks of northeastern China, the BIF layers are generally associated with amphibolites and mafic granulites interpreted to be metabasalts. For this reason, Zhang (1987) and Sills et al. (1987b) equate the Archean BIF-metabasalt association of China with the Algoma-type. The early Proterozoic BIFs in northeastern China are comparable to the Superior-type because they lack a major volcanic association, but in northeastern China these Superior-type BIFs are much less extensive than in North America, and most of the iron ore comes from the Archean Algoma-type BIF.

The iron formations in northeastern China are almost exclusively of the oxide facies (dominated by magnetite), with subordinate iron silicates. One reason for the lack of carbonate-facies iron formation and for the relatively low oxidation state of the Chinese BIF may be the high-grade metamorphism which the rocks have undergone. Lower crustal oxygen fugacities are typically below the magnetite-hematite buffer, and carbonate minerals in association with silica would be destroyed by decarbonation reactions during prograde metamorphism such as the following:

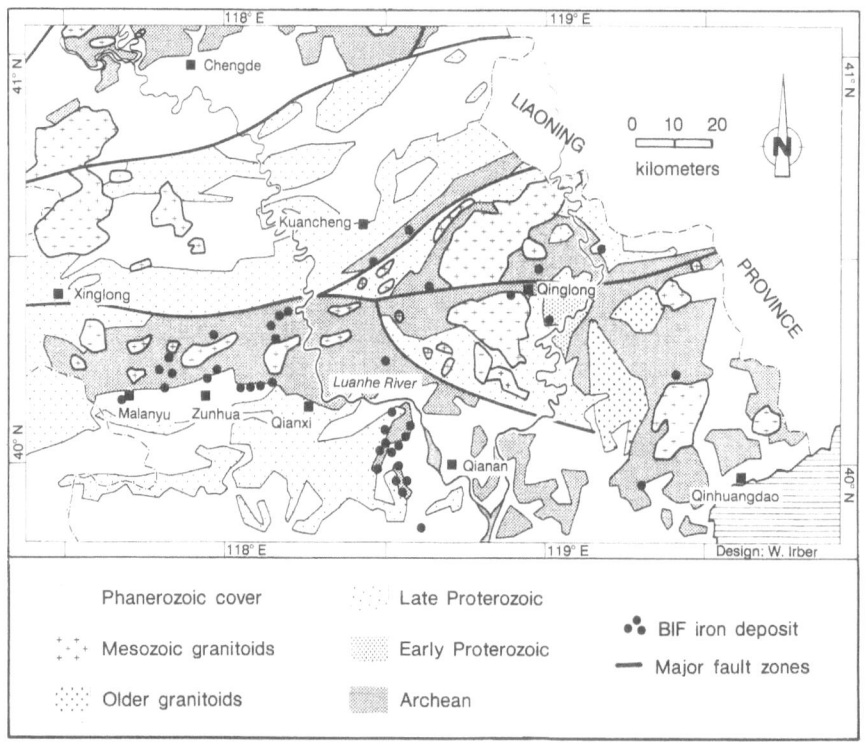

Fig. 4.6. Simplified geologic map of part of eastern Hebei province showing the distribution of BIF-type iron deposits

$$Ca(Fe, Mg)(CO_3)_2 + 2SiO_2 = Ca(Fe, Mg)Si_2O_6 + 2CO_2.$$

$$(Fe, Mg)CO_3 + SiO_2 = (Fe, Mg)SiO_3 + CO_2.$$

Turning our attention to eastern Hebei province, BIF occurs as discontinuous, stratiform lenses in the Archean Qianxi and Dantazi Groups, and to a lesser extent in the Early Proterozoic Zhuzhangzi Group. Figure 4.6 shows the distribution of the largest iron deposits in this area. The comparison of this map with the distribution of gold deposits in the same area shown on Fig. 1.3 indicates a poor correspondence of iron and gold deposits, a fact also noted by Shen et al. (1989). The largest iron deposits in eastern Hebei province are found in the Archean Santunying Formation in the Qianan area south of the Luanhe River. In this area, BIF layers are continuous along strike for up to several kilometers and reach a thickness of up to 100 m. These BIFs have been described by Zhai and Windley (1990). The BIF layers north of the Luanhe River, where almost all of the gold districts are located, occur in amphibolites and gneisses as discontinuous, stratiform, mostly lens-shaped bodies of minor economic importance. The observed thickness of the iron-rich lenses in this area is mostly some decimeters to several meters, and their lateral extent is some tens to hundreds of meters, depending on local structure. Banding of quartz-rich and magnetite-rich portions in the iron-rich rocks is locally present, but it is by no means ubiquitous, and the term banded iron formation is often not justified. In the Chinese literature the iron-rich rocks are commonly called magnetite quartzites and this term will also be used here.

4.3.2 Petrography and Composition of Magnetite Quartzites

The magnetite quartzites of the Qianxi Group exposed in the gold districts north of the Luanhe River grade continuously over a distance of a few decimeters into normal amphibolites. In the most iron-rich layers magnetite and quartz occur in approximately equal proportions, and hornblende and plagioclase are reduced to accessories. Such rocks possess a foliation and banding which is continuous with that of the surrounding amphibolites.

Like their amphibolite country rocks, the magnetite-rich rocks have been metamorphosed to granulite or upper amphibolite facies conditions and show abundant evidence of later retrograde metamorphism in the form of fibrous blue-green actinolite pseudomorphs after pyroxene and hornblende, chlorite after garnet, and sericitization of amphibole and plagioclase. The rocks have a hypidiomorphic granular texture with weak foliation, and are locally banded, with quartz-rich and magnetite-rich layers. Magnetite in the rocks is intergrown with quartz and/or retrograde fibrous amphibole. The magnetite has a grain size of about 0.2 to 1 mm.

Chemical analyses of selected samples of magnetite quartzites collected from areas north of the Luanhe River are shown in Table 4.3. Some samples were taken from outside the gold mining districts and others from outcrops in the mineralized areas of the Niuxinshan and Sanjia districts. The table is divided into two sections. In section (a) those samples are listed which show little or no secondary alteration (based on petrography and total S content). Table 4.3b lists analyses of samples affected by secondary sulfidization. The latter are discussed separately in the following section. The total iron concentrations of unaltered samples (as Fe_2O_3) range from 28 to 47 wt%. The concentrations of MgO and CaO range from about 1 to 2 wt%. The range of Al_2O_3 is from less than 1 to about 3 wt%. The trace element concentrations are all quite low with the exception of chalcophile elements in some samples affected by sulfidization as discussed below. These chemical compositions compare favorably with those given in Zhai and Windley (1990) for BIF ores from various parts of northeastern China except that the total iron contents reported here are lower, and the Al_2O_3 concentrations are higher than the range reported by Zhai and Windley (1990). This discrepancy is probably due to the fact that the samples reported here were

Table 4.3. Partial chemical analyses of magnetite quartzite samples from the Qianxi Group

| Locality[a] | a (Unaltered) | | | | | b (Sulfidized) | | | | | | | |
Sample No.	NXS 6126	SJ 6203	STY 6101	NXS 6166	NXS 6122	NXS 6128	WJ 6321	STY 6093	STY 6095	NXS 6170	NXS 6185	NXS 6523	STY 6102
TiO_2	0.09	0.22	0.04	0.41	0.63	1.55	0.01	0.16	0.06	0.37	0.01	0.05	0.32
Al_2O_3	1.24	6.10	1.37	1.17	0.96	12.29	0.23	2.58	1.98	4.20	0.55	3.77	1.14
Fe_2O_3	49.01	26.03	33.20	45.42	47.06	14.43	45.42	25.61	37.31	28.32	42.30	32.66	32.70
Mn	449	534	527	774	790	1626	372	457	534	984	4848	7977	1967
MgO	1.18	1.90	2.18	1.34	0.85	5.41	0.55	1.86	1.38	3.23	1.94	1.22	0.77
CaO	1.35	3.20	2.90	1.69	0.81	6.21	1.87	2.51	3.01	3.64	1.75	1.52	2.59
Na_2O	0.13	0.89	0.05	0.06	0.03	3.57	0.01	0.22	0.04	0.74	0.01	0.02	0.02
K_2O	0.13	1.90	0.00	0.04	0.35	1.22	0.00	1.48	0.25	0.25	0.19	0.56	0.05
P_2O_5	0.25	0.25	0.21	0.23	0.23	0.13	0.28	0.58	0.22	0.19	0.22	0.13	0.20
Rb	25	99	25	25	60	144	25	104	56	25	57	141	25
Sr	59	300	34	11	13	337	49	99	55	73	48	28	25
Ba	48	1190	19	12	59	522	25	355	59	31	39	58	21
V	1	24	1	5	55	314	1	21	3	44	1	9	26
Cr	3	18	1	1	1	42	2	31	3	40	1	5	1
Co	15	9	6	15	8	38	1	3	1	37	1	14	21
Ni	6	18	8	20	8	35	1	24	5	96	1	28	28
Y	9	12	8	8	6	30	8	20	8	11	6	7	8
Zr	1	14	12	1	1	7	1	2	15	4	1	9	1
Total S (wt%)	0.01	0.01	0.01	0.04	0.05	0.10	0.51	0.71	0.95	2.64	2.72	7.45	14.27
Cu	11	5	6	10	29	58	29	114	100	644	209	1195	844
Zn	57	66	36	113	115	158	21	29	120	81	277	384	214
Pb	32	5	5	5	5	34	5	15	5	5	49	221	14
Ag	0.8	0.6	0.6	0.6	0.9	0.6	0.3	1.2	1.1	1.0	1.5	5.0	3.0
W	27	5	16	15	36	5	17	25	5	15	33	306	36
Au (ppb)	1	33	1	35	17	287	2	33	299	28	87	461	4590

Oxides in wt% elements in ppm unless otherwise stated, total Fe as Fe_2O_3.
[a] Localities: NXS, Niuxinshan; SJ, Sanjia; WJ, Wangjiangzi; STY, Santunying.

not taken from iron ore but represent the typical subeconomic magnetite quartzite horizons in the Qianxi Group.

4.3.3 Sulfidization and Gold Mineralization

The gold concentrations in the non-sulfidized samples of magnetite quartzite in Table 4.3a are at or below the Clarke average, and they are comparable to those of the Qianxi Group amphibolites in general (Fig. 4.2, Appendix 2). This suggests that the iron formations themselves are not primarily enriched in gold. Any gold enrichment noted in the samples from the mining districts can be attributed to secondary sulfidization.

Petrography of Sulfidization
Sulfidization is manifested by the replacement of magnetite by sulfide minerals including mostly pyrite and lesser pyrrhotite and chalcopyrite. Typical of the newly formed pyrite grains are numerous minute inclusions of silicate minerals which were present as inclusions and interstices within magnetite aggregates. The sulfidization front can be seen in some samples as a clearly delineated zone which forms around a central quartz vein. An example is shown in the photomicrographs in Fig. 4.7. The ore minerals found in samples of sulfidized magnetite quartzite include: magnetite, pyrite, pyrrhotite, chalcopyrite, native gold. They show the following paragenetic features:

Magnetite
Magnetite is usually found in aggregates of up to a few millimeters in size within or in the interstices of the silicate minerals. The magnetite enrichment occurs in bands which parallel the foliation of the rock. The magnetite grains are euhedral to subhedral, and commonly fractured and partially replaced by Fe-hydroxides, chlorite, carbonates, and sulfides.

Pyrite and Chalcopyrite
Pyrite occurs in the magnetite quartzite within the selvage zone of quartz-sulfide veins as cubes of some millimeters in size and as irregularly shaped aggregates pseudomorphous after magnetite. Pyrite replaces magnetite along fractures and grain boundaries. The pyrite pseudomorphs after magnetite contain numerous inclusions of silicates, chalcopyrite, pyrrhotite, and relict magnetite grains. An example is shown in Fig. 4.7.
Chalcopyrite forms irregular aggregates of sub-millimeter size at the replacement front between magnetite and pyrite. It can also be observed to fill cracks in cataclastic pyrite and it occurs in minor amounts disseminated as small anhedral grains associated with mafic silicates.

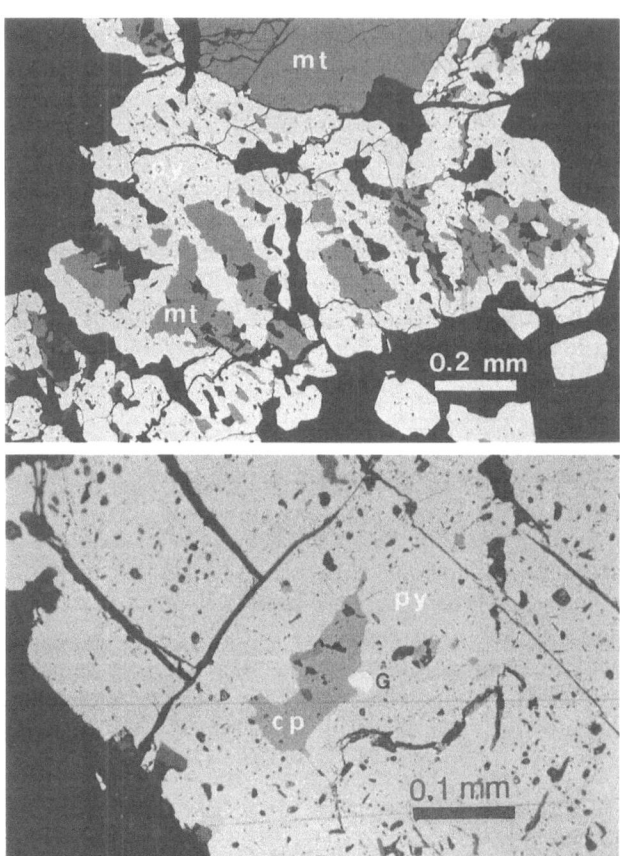

Fig. 4.7. *Above*: Photomicrograph of nearly complete replacement of magnetite (*mt*) by pyrite (*py*) in sulfidized magnetite quartzite from the Sanjia deposit. *Below*: Gold (*G*) and chalcopyrite (*cp*) in pyrite (*py*) in sulfidized magnetite quartzite from the Sanjia deposit

Native Gold

Gold is quite rare in the sulfidized magnetite quartzites. Single, free gold grains up to about 0.01 mm in size occur within pyrite grains along the sulfidization front. An example is shown in Fig. 4.7 (below). The gold grains may be intergrown with chalcopyrite. The EDS microanalysis of gold (one sample) showed 25 wt% Ag, which is not appreciably different from the composition of gold from ore in the normal amphibolite host rocks.

Geochemistry of Sulfidization

The degree of alteration of the samples is reflected in their chemical compositions by variations in the K_2O and total S concentrations (cf. Table 4.3a

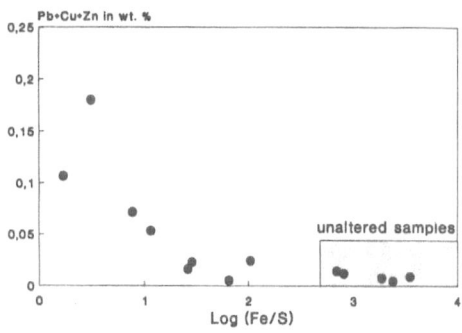

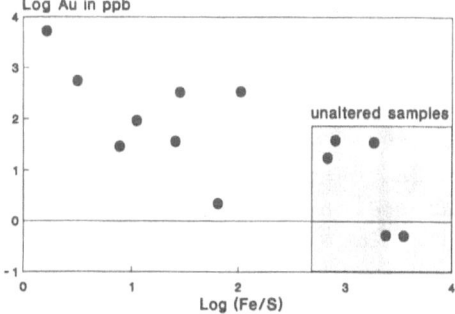

Fig. 4.8. Diagrams showing the variation of base metals concentrations (*upper*) and gold (*lower*) with the degree of sulfidization in magnetite quartzites from the Qianxi Group

and b). Along with the increase in sulfur, the chalcophile elements Cu, Pb, Zn, and Au are also enriched relative to the unaltered magnetite quartzites. The effect of sulfidization on gold and base metal concentration is examined in Fig. 4.8, which shows the element concentrations plotted against the weight ratio of Fe/S as a measure of the sulfidization degree. The increase in total base metal concentration of the rocks with sulfidation is clearly demonstrated. An increase in gold concentration with degree of sulfidization is apparent but the trend is much less regular.

Precipitation Mechanisms

According to the present knowledge of gold geochemistry (Seward 1991), gold will be present as both chloride- and sulfur complexes at the physical conditions of ore formation discussed in Chapter 4.6 (ca. 300–350°C and 2–4 kbar). The bisulfide complex of gold $[Au(HS)_2^-]$ can be de-stabilized by sulfidization of magnetite during wall rock alteration because these reactions remove reduced sulfur from the hydrothermal fluid (Neall 1987). An overall model reaction for this process is:

$$6Au(HS)_2^- + 2Fe_3O_4 + 0.5O_2 = 6Au° + 6FeS_2 + 3H_2O + 6OH^-.$$

Note that the formation of pyrite from iron-rich silicate minerals would have the same effect of consuming sulfide from the fluid. Phillips and Groves (1984) proposed the following model reaction for the replacement of iron-rich chlorite by pyrite and magnesian chlorite:

$$3Fe_4Mg_2Si_2O_{10}(OH)_8 + 12H_2S + 3O_2 =$$
$$6FeS_2 + 2Fe_3Mg_3Si_4O_{10}(OH)_8 + 4SiO_2 + 16H_2O.$$

4.3.4 Conclusions

There is no evidence that the iron-rich rocks (magnetite quartzites) in the gold districts of eastern Hebei province have a primary, syngenetic gold enrichment. Instead, the high gold concentrations measured in some samples of these rocks are a result of secondary sulfidization of the rocks caused by hydrothermal alteration. The reactions forming pyrite from magnetite or from iron-rich silicate minerals should favor precipitation of gold carried in solution as a sulfur complex. The observed positive correlation of gold concentration with degree of sulfidization is good evidence that sulfur complexes were in fact involved in the transport of at least some of the gold present in these rocks.

At present there is no direct confirmation from mine records that magnetite quartzites and/or iron-rich horizons in the host rocks are particularly well mineralized with gold. On the contrary, it was noted above that the large iron deposits in eastern Hebei province are not notably mineralized with gold (see also Shen et al. 1989). However, the iron deposits do not occur in exactly the same geologic setting as the gold deposits, and in the gold mining districts themselves this aspect of host-rock control of mineralization has not previously been considered. A reexamination of drill core logs in this light might give valuable information for further gold exploration. If a connection is found, tracing the magnetite quartzite layers in the field by inexpensive magnetometry would be an effective exploration tool.

4.4 The Role of Yanshanian Granites

The field relations in the individual mining districts of eastern Hebei province discussed in Chapter 3 and the evidence from isotopic dating summarized in Chapter 4.1 clearly demonstrate that gold mineralization is closely related in time and space with certain Yanshanian granite plutons and associated dikes. In most districts studied, the ore veins cut the associated granites and are penecontemporaneous with post-granite mafic and felsic dikes. Some field relations are ambiguous, but in general it can be stated that the mineralization occurred in a late stage of Yanshanian magmatism. Other workers have also proposed that the Precambrian-hosted gold deposits in northeastern China are genetically related to Yanshanian granites (M.Z. Yang 1988; Liu 1989; Zhou 1989), and the same conclusion was reached by Shelton et al. (1988) in a study of similar Archean-hosted gold deposits in South Korea.

Of course, it is well known that the Yanshanian magmatism (in the broad sense, Mesozoic–Tertiary, see Chap. 2.4) was important to the formation of Sn, W, Cu, and Mo ore deposits in eastern Asia generally. Well-documented areas include southeastern Asia (Hutchinson and Taylor 1978), southeastern China (Guo 1982; Ishihara and Sato 1982; Wu 1985), southern Korea (Lee 1981; Iyama and Fonteilles 1981; Kim and Lee 1983), and Japan (Ishihara 1981; Ishihara et al. 1981). The information from northeastern China and Siberia is less complete, but in these areas, too, a strong correlation of ore deposits with Yanshanian magmatism has been noted (Zonenshain et al. 1974; Guo 1982; Ishihara 1984).

Despite this supporting evidence, the connection between granitic intrusions and gold mineralization is not unambiguous, and the role of granites in the genesis of gold deposits is not so well understood as in the case of Sn, W, Cu, and Mo deposits. It is important to note, too, that granite-related gold metallogenesis in northeastern China is not entirely restricted to the Yanshanian Orogeny. In some gold mining districts in Jilin and Heilongjiang province discussed in Chapter 1.2, the granites associated with gold deposits are Variscan in age.

4.4.1 Granite Compositions

The spatial distribution of Yanshanian granites and gold deposits in eastern Hebci province (Fig. 1.3) raises the question of why certain granites are associated with gold and others are not. This is of obvious strategic importance to further exploration, and one important aspect of the problem is the composition of the various plutons. Unfortunately, there have been relatively few studies done on the Yanshanian granites in northeastern China, and there are no comprehensive data with which to compare the granites associated with gold deposits and those with no gold association. The authors began such a study in eastern Hebei province with a geo-chemical reconnaisance of 16 Yanshanian plutons as presented in Chapter 2.4.3. The results of this study which are relevant to the granite–gold relationship are discussed below.

According to the work of Ishihara (1984) and Wu (1985) on granite-related metallogeny in eastern Asia as a whole, it might be expected that granites associated with gold deposits would be intermediate in composition and belong to the magnetite series, in contrast to the ilmenite-series leucogranites associated with Sn and W deposits. This idea is supported by Tu (1989), who stated that the granitoids associated with gold deposits in China tend to have "dioritic" compositions. M.Z. Yang (1988) proposed three genetic categories of granites associated with gold deposits in northern China based on their geologic setting, initial $^{87}Sr/^{86}Sr$ and oxygen isotope ratios. The categories are: anatectic crustal source granites, deep-source granites, and a transitional type termed strata-bound deep source granites. Gold deposits occur with granites of all three categories according to M.Z. Yang (1988), and he gives no indication that any one granite type is more important than the others for gold metallogenesis. It may also be relevant to this discussion that the magmatogenic Late Tertiary gold deposits in western Pacific island arcs are associated with both rhyolitic-rhyodacitic rocks and andesitic-dacitic rocks, although the latter predominate (Sillitoe 1989).

The compositions of 16 Yanshanian granites from eastern Hebei province are summarized in Appendix 1 and were discussed in Chapter 2.4.3. Two compositional groups were recognized, the dominant one characterized by relatively felsic, moderately differentiated granites; and a second group containing only a few plutons which are more mafic and weakly differentiated. Table 4.4 and Figs. 4.9 and 4.10 show selected geochemical features of the plutons grouped according to the presence or absence of gold deposits

Table 4.4. Selected chemical features of gold-related and not-gold-related Yanshanian granites in eastern Hebei province

Pluton	SiO_2	Fe index	Rb/Sr	K/Rb	Na_2O/K_2O
Gold-related granites					
Laochengling	76.66	0.63	–	–	1.13
Sanyihe	73.93	0.49	1.5	245.7	1.07
Madi	74.15	0.24	367.6	18.9	1.50
Yuerya	73.97	0.32	0.9	260.0	0.81
Maoshan	74.46	0.33	1.4	96.4	1.03
Niuxinshan	74.25	0.57	10.0	144.6	0.65
Qingshankou	73.57	0.37	1.4	235.1	0.99
Wangtoushan	75.69	0.51	0.9	237.7	0.62
Gaojiadian	56.30	0.40	0.1	492.2	1.56
Not-gold-related granites					
Xiaoyingzi	68.79	0.47	0.1	590.1	1.03
Wangpingshi	73.36	0.44	1.7	159.3	0.94
Doushan	70.07	0.34	0.1	512.0	0.99
Luowenyu	75.35	0.28	5.4	124.1	1.02
Qianfenshuiling	69.76	0.52	–	–	1.29
Jiajiashan	73.93	0.38	2.1	223.9	1.16
Shitaizi	73.52	0.59	–	–	0.81

Fe index: atomic ratio $Fe_2O_3/(Fe_2O_3 + FeO)$.
A value of 0.35 divides the ilmenite and magnetite series of Ishihara (1984).

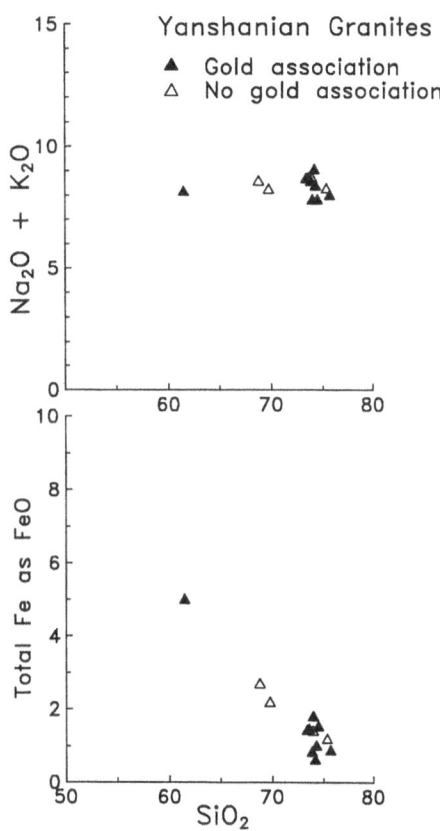

Fig. 4.9. Harker diagrams showing the average composition of Yanshanian plutons in eastern Hebei province with spatially associated gold deposits and plutons with no gold association

spatially associated with them (gold-related and not gold-related plutons). The average compositions of the plutons were used to avoid overcrowding the diagrams and to give equal weight to plutons from which many samples were analyzed and those represented by only one or two analyses. The data show that the gold-related and the not-gold-related plutons both cover a wide range of compositions, and that these ranges overlap. The two compositional groups of Yanshanian plutons identified in Chapter 2.4.3 do not correlate with the presence or absence of gold mineralization. The gold-related plutons in Table 4.4 belong mostly to Ishihara's (1984) magnetite series according to their whole-rock oxidation index (atomic $Fe_2O_3/Fe_2O_3 + FeO > 0.35$), but this does not distinguish them from the plutons not related to gold deposits.

All available data on gold concentrations in Yanshanian granites from eastern Hebei province were presented in Table 4.2. It suffices to reiterate

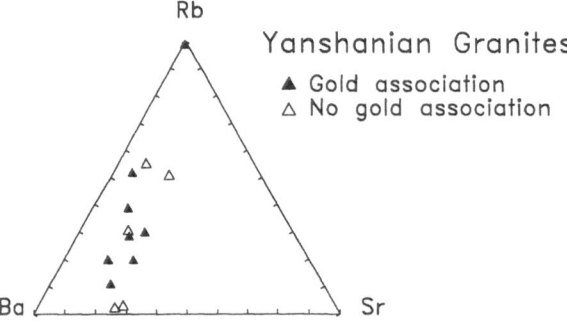

Fig. 4.10. Rb-Ba-Sr diagram showing the average composition of Yanshanian plutons in eastern Hebei province with spatially associated gold deposits and plutons with no gold association

here that the granites associated with gold deposits have "normal" background gold concentrations at or below 5 ppb. There are not enough data at present to compare the gold concentrations in gold-related and in not-gold-related plutons, and this aspect should be pursued in future studies.

4.4.2 Heat Production

Recently, the concept of high heat production granites was proposed as an important factor in granite-related metallogenesis. A high heat production granite is one with a high concentration of radioactive elements (mainly K, Th, and U), which constitute a significant heat source long after the magmatic heat has dissipated. The consequence is that hydrothermal circulation around such granites can last much longer than otherwise, and this factor alone may increase the mineralization potential of such granites. It is thought that, for example, the hydrothermal ores associated with the Cornubian batholith of southwestern England were produced or enhanced by this process, and even today the area exhibits anomalously high surface heat flow and a high geothermal gradient (Stone and Exley 1985).

The heat production of a granite can be calculated from a chemical analysis, and the results for selected plutons in the mining districts in eastern Hebei province are shown in Table 4.5. Unfortunately, the requisite U and Th data are presently available for only three granites, and all three are associated with gold deposits. The heat production value of the Niuxinshan granite is comparable with that of the Cornubian batholith, which may be considered the "type" high-heat production granite, but the other two Chinese granites have lower heat production values than even the average granite. Therefore, it seems that there are both "normal" granites and granites of the high-heat production type in eastern Hebei province. However, more data are

Table 4.5. Radiogenic heat production of selected granites from eastern Hebei province

Pluton	U (ppm)	Th (ppm)	K_2O (%)	Heat production (10^{-6} W/m^3)
Niuxinshan	7.6	32	4.79	4.66
Sanyihe	2.7	11.3	4.50	1.88
Wangtoushan	3.1	7.0	5.15	1.72
Average granite[a]	3	17	5.06	2.49
Cornubian batholith[b]	11.3	19.1	4.3	4.7

Heat production calculated from average composition of the granites using the equation of Rybach (1976) assuming a density of 2.7 for all granites.
[a] Compositional data from Rösler and Lange (1972).
[b] Data from Brown et al. (1979).

needed, including analyses from plutons not related to gold deposits, before the significance of this distinction can be discussed.

4.4.3 Conclusions

The field relationships and radiometric age data demonstrate that gold mineralization in eastern Hebei province is related in space and in time to certain of the Yanshanian granites and/or their related igneous dikes. However, the relation is not a direct one; that is, the granites themselves are seldom mineralized and in most cases they were in place and solid before mineralization commenced.

The chemical composition of 16 Yanshanian granites from eastern Hebei province, including both gold-related and not-gold-related plutons, and the lead isotopic composition of ore minerals from the deposits (Chap. 4.2.3) suggest the following conclusions:

1. Apart from local high values due to alteration, the gold concentrations (<1 to 5 ppb) of the gold-related granites (i.e., Niuxinshan, Wangtoushan, Sanyihe, Maoshan, Yuerya) are not above average, and the concentration of gold in these granites is lower than or equal to that in the Archean country rocks.
2. The lead isotopic compositions of galena and iron sulfides from the ore veins suggest that the lead in the gold deposits was derived from the Early Precambrian metamorphic basement and not from the Yanshanian granites.
3. The compositions of individual granites in eastern Hebei province do not correlate with the presence or absence of associated gold mineralization. Gold deposits are associated with granites having a wide range of composition.

4. Based on whole-rock Fe_2O_3 and FeO concentrations, both the granites associated with gold deposits and those with no gold association belong predominantly to the magnetite series of Ishihara (1984).
5. Both low- and high-heat production granites occur in the gold mining districts of eastern Hebei province. More studies are needed to assess the importance of radiogenic heat production to mineralization potential in this area.

In summary, the available evidence suggests that the chemical composition of the Yanshanian granites has no bearing on the presence or absence of gold mineralization, and that the granites were not the source of ore-forming elements found in the deposits. However, the close spatial and temporal relationships of Yanshanian granites and associated dikes with gold mineralization in all districts studied cannot be coincidental. If the granites made no material contribution to the deposits, the only reasonable explanation for the relationship is that the heat and/or deformation attending the intrusion of granites created conditions favorable for hydrothermal circulation. The nature of faults and fractures in the gold districts would play a vital role in focusing or dispersing hydrothermal circulation, and it may be that the difference between gold-related and not-gold-related granites lies in the local structural environments, and has nothing to do with the composition of the granites themselves.

4.5 Structural Controls

Rock structures, in particular fault zones and fractures, are the most direct factors controlling the form, distribution and intensity of gold mineralization in eastern Hebei province on both a regional and a local scale. This section summarizes the most important features of structure relevant to gold mineralization based on the information given in Chapter 2.3 (regional structure) and Chapter 3 (local structures in the mining districts). We then attempt to interpret these features in relation to regional tectonic movements in northeastern China, especially those of the Yanshanian Orogeny.

4.5.1 Regional Structures

On a regional scale, the distribution of gold mining districts in eastern Hebei province and surrounding areas is closely bound to uplifts of Early Precambrian basement rocks. Within these, the mining districts are associated with major fault zones. The structural fabric of the metamorphic basement rocks (i.e., folds and foliation) has little direct influence on the distribution of gold deposits, although it may locally exert an indirect influence inasmuch as the locus of faulting and igneous intrusion in some districts (e.g., Niuxinshan) coincides with the hinge lines of major folds.

The regional-scale fault zones in eastern Hebei province and surrounding areas form two main groups, striking roughly E–W and NE–SW respectively, and a third, less well-developed group with NW–SE strike direction (Fig. 2.6). Most of the faults formed in a compressive tectonic regime and show reverse or oblique-slip senses of offset. In general, the E–W-trending faults are earlier than, and are offset by, the NE–SW faults. The NW–SE-trending faults cross-cut both of the other fault groups and are therefore later. Both the E–W and NE–SW groups of faults show evidence of movement beginning in the Early Proterozoic, with episodes of later reactivation. However, the most active period of faulting in northeastern China was during the Mesozoic Yanshanian Orogeny (Ren et al. 1987). Much of the Yanshanian faulting involved a reactivation of earlier structures, especially those oriented NE–SW. In addition, faults with NE–SW and NW–SE trends were newly formed.

The gold districts in eastern Hebei province and surroundings are located within a few kilometers of the major fault zones, and they are especially common near the intersections of E–W- and NE–SW-trending fault zones. Individual ore deposits occur along subsidiary faults, and the major faults themselves are barren. In this sense, the deposits in eastern Hebei province are similar to many Archean mesothermal gold deposits in Western Australia (Eisenlohr et al. 1989) and Canada (Colvine 1989). A general discussion of possible factors relating gold mineralization to deformation processes in shear zones is given by Eisenlohr et al. (1989).

4.5.2 Local Structures

The gold mineralization on the scale of mining districts or single deposits is strictly controlled by faults and fracture zones. Most of the ore-controlling faults form brittle or brittle-ductile shear zones a few decimeters to meters wide which show reverse or oblique-slip sense of offset. The ore veins have a pinch-and-swell structure and the vein-filling took place partly during shear, as is shown by schlieren of phyllosilicates and strained, recrystallized quartz subgrains oriented parallel to the vein borders. Local dilation of the veins is evidenced by laminar structure produced by multiple episodes of opening and quartz filling (crack-seal mechanism).

The dilation responsible for vein-filling can be explained by a number of possible factors (Harris 1987): (1) secondary dilational structures within major shear zones, (2) irregularities along faults due to different rock competency (pull-apart zones), (3) intersections with zones of preexisting weakness. Factors which could cause district-wide dilation of pre-existing faults include: (1) changes in the regional stress orientation, and (2) uplift caused by granite intrusion.

Structural studies of ore-bearing veins in the Niuxinshan and Sanjia districts by the Ministry of Metallurgical Industry suggest that rotation of the regional stress field and local uplift due to granite intrusion were the main causes of vein-filling. No detailed studies of ore field structures have been made in the other districts. Zhai (1984) emphasized the importance of changes in the regional stress orientation with time as a metallogenetic factor in many vein-type ore deposits in northeastern China.

4.5.3 Regional Tectonic Interpretation

It is well known from regional studies that the main orogenic compressional direction in northeastern China has changed with time (Ren et al. 1987), and this rotation of the regional stress field with time forms the basis of a tectonic interpretation linking local structures in the mining districts with regional orogenic events.

From the Early Proterozoic until the Mesozoic, the main direction of orogenic compression in northeastern China was N–S, related to convergence along the northern and southern margins of the Sino-Korea platform. This compression produced E–W-trending fold axes which refolded Early Archean structures and gave rise to the present NE- and SE-trending structural grain of the basement. Related Proterozoic strike-slip or reverse faults trend E–W or ENE–WSW.

At the end of the Triassic period, convergence of the Izanagi plate with the Sino-Korean Platform began on the Pacific margin and caused a rotation of compression to ESE–WNW. However, the N–S direction of compressional stress still dominated. In the course of the first and second stages of the Yanshanian Orogeny (late Jurassic and Early Cretaceous) the main compression direction rotated from N–S to ESE–WNW. NNE-striking faults resulted, and open folds formed in Jurassic strata with NNE-striking axial planes. During the Cretaceous and Early Tertiary (third Yanshanian stage) the direction of main compression shifted to NNE–SSW (Wan and Zhu 1989).

A comparison of the time-dependent regional stress orientation summarized above and the orientation of local principal stress directions inferred from structural analysis in the mining districts is shown in Fig. 4.11. For each mining district in which structural studies were done, the figure indicates the orientation of maximum bulk compressive stress (σ_1) for the periods of time before, during, and after mineralization. The orientations of the principle stress axes are based on field measurements of inferred conjugate shear planes showing the sense of movement and a clear age relationship to ore veins. Note that all measured structures, including those predating the mineralization, are post metamorphic, brittle structures. There are slight differences in the plunge of the maximum principle stress direction from one mining district to another but all show a very similar SE–NW orientation. This orientation fits with the regional stress field at the time of the first stage of the Yanshanian Orogeny. It is inconsistent with the tectonic regimes of

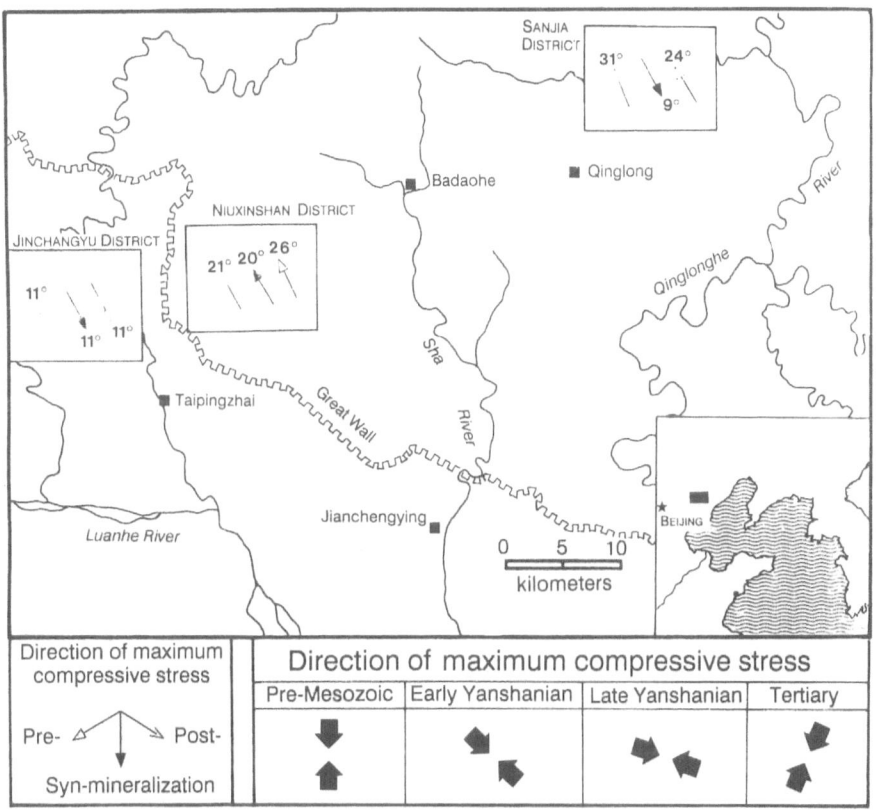

Fig. 4.11. Inferred directions of maximum compressional stress derived from pre-, syn-, and postmineralization structures in selected gold mining districts in eastern Hebei province. See text for explanation. The *inset* shows the directions of maximum compression during the main orogenies affecting northeastern China

both the pre-Mesozoic orogenies and the Cretaceous/Early Tertiary third stage of the Yanshanian Orogeny. In fact, even the post-mineralization structures show no evidence of the NNE–SSW Cretaceous compressional direction.

In summary, the orientation and sense of movement of the ore-controlling structures in the gold districts of eastern Hebei province suggest that the main period of mineralization was during the first stage of the Yanshanian Orogeny. This age constraint agrees with the radiometric dates obtained from the vein minerals and with the relations of mineralization with granites and dike intrusions of Early Yanshanian age.

4.6 Fluid Composition and P-T Conditions of Mineralization

4.6.1 Evidence from Fluid Inclusions

Table 4.6 gives a summary of the fluid inclusion characteristics from gold deposits in eastern Hebei province for which data is available. The data from the Jinchangyu and Yuerya deposits were taken from Yu and Jia (1989) and Yu et al. (1989); the data on the other deposits are from investigations carried out in the course of this project. More detailed descriptions of the inclusions are given for each deposit separately in the respective sections of Chapter 3. The microthermometric data in Table 4.6 refer to the primary inclusions only. The compositional data are based on chemical analyses of bulk inclusion contents which unavoidably represent mixtures of primary and secondary inclusions. The contribution of the latter was minimized by analyzing only the fluid released after most secondary inclusions had decrepitated. Nevertheless, the compositional data should be interpreted with caution.

Table 4.6. Summary of fluid inclusion characteristics and composition from selected gold deposits of eastern Hebei Province

Deposit name	Inclusion characteristics
Niuxinshan	*Microthermometric data* H_2O-CO_2, 3-phase inclusions at room temperature Salinity: 6–9 wt% NaCl equivalent, 9–12 mol% CO_2 Final homogenization T: 270–315 °C Bulk fluid density: 0.9–0.96 *Fluid composition from bulk analysis, average mole fraction* XNa: 0.72 XK: 0.10 XCa: 0.12 XMg: 0.05 XCO_2: 0.051
Wangtoushan	*Microthermometric data* H_2O-CO_2, 3-phase inclusions at room temperature Salinity: 6–8 wt% NaCl equivalent, 10–14 mol% CO_2 Final homogenization T: 190–340 °C Bulk fluid density: 0.91–1.01 *Fluid composition from bulk analysis, average mole fraction* XNa: 0.72 XK: 0.15 XCa: 0.10 XMg: 0.03 XCO_2: 0.046
Yuerya	*Microthermometric data* H_2O-CO_2, 2- and 3-phase inclusions at room temperature Final homogenization T: 260–390 °C *Fluid composition from bulk analysis, average mole fraction* XNa: 0.66 XK: 0.33 XCa: 0.01 XMg: 0.01 XCO_2: 0.260
Jinchangyu	*Microthermometric data* H_2O-CO_2, 2- and 3-phase inclusions at room temperature Final homogenization T: 256–370 °C *Fluid composition from bulk analysis, average mole fraction* XNa: 0.58 XK: 0.16 XCa: 0.21 XMg: 0.06 XCO_2: 0.030

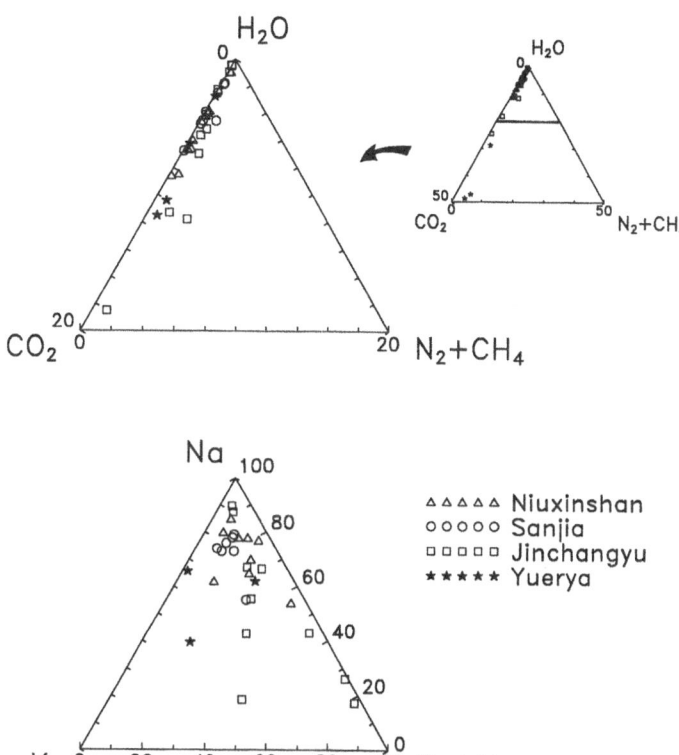

Fig. 4.12. Summary of bulk fluid composition from selected gold deposits in eastern Hebei province

Figure 4.12 summarizes the compositional data from the inclusion studies. Subject to the caveat given above, the similarity of fluid characteristics in these deposits is remarkable. In all of the investigated deposits, the inclusion fluids are dominated by H_2O with typically about 10 mol% of CO_2 (based on microthermometry and visual volume estimates), and they contain minor but detectable quantities of CH_4 and N_2 (based on chemical analysis of inclusion contents). Microthermometric data indicate low salinities, from 6 to 8 wt% NaCl equivalent, and bulk densities of 0.9 to 0.95 g/cm^3. Chemical analyses of bulk inclusion contents indicate that the dissolved cations are dominated by Na, but K, Ca, and Mg are also important constituents.

It may be significant that the fluid inclusion compositions from amphibolite-hosted deposits closest to Yanshanian granites, in which the alteration zones contain K-feldspar and fluorite (Niuxinshan and Wangtoushan), are no different from those where the nearest granite is a few kilometers distant (Jinchangyu and Sanjia), suggesting that a granitic fluid input in these deposits was minimal. The inclusions from the Yuerya deposit are rich in

CO_2 and K compared with those from the other deposits. This difference may be due to the host rocks. The Yuerya deposit is hosted in a potassic granite near the contacts with carbonate rocks, whereas all of the other deposits are hosted in amphibolites, which have much higher whole-rock Na/K and Na/Ca + Mg ratios than the host rocks at Yuerya.

The final homogenization temperatures of the primary inclusions measured range from 190–390 °C, but the inclusions in most deposits show a clustering of homogenization temperatures in the narrow range of 250–350 °C (Table 4.6). These temperatures are not pressure-corrected and therefore represent the minimum estimates of mineralization temperature. The minimum pressure, based on the CO_2-H_2O solvus with 6 wt% NaCl according to Brown and Lamb (1989), is about 2–4 kbar for the range 250–350 °C. There is no evidence of heterogeneous trapping in the inclusions studied, and it is therefore likely that the trapping temperatures were higher than the homogenization temperatures recorded. An independent estimate of the mineralization temperature is given by oxygen isotope thermometry of hydrothermal quartz and K-feldspar pairs. The temperatures from the Niuxinshan deposit are 430–490 °C (see Chap. 3.1.7), and the intersection of this temperature range with the fluid inclusion isochores indicates model pressures for mineralization of 3.5 to 5 kbar. These estimates seem high considering the brittle nature of the mineralized veins. Furthermore, one of the four samples analyzed for oxygen isotope thermometry from the Niuxinshan deposit gave an impossibly high temperature of cover 1000 °C, and the oxygen isotopic data from the Wangtoushan deposit showed isotopic disequilibrium. Therefore the results of isotopic thermometry are of questionable validity, and the microthermometric data are considered to provide the most reliable estimates for the P–T conditions of mineralization.

The type of fluid inclusions found in these deposits and the range of homogenization temperatures recorded are similar to those reported for mesothermal gold deposits from Archean terranes elsewhere (Ho et al. 1990; Groves and Foster 1991), but also for metamorphic-hosted lode gold deposits of Phanerozoic age (Nesbitt 1991). As discussed by these authors, such low-salinity, CO_2-rich fluids are not diagnostic of any one fluid regime. Possibilities include metamorphic devolatilization, but in the case of eastern Hebei province, this hypothesis (metamorphic fluids) can be disregarded since the gold mineralization took place long after the last phase of regional metamorphism. Fluid inclusions typical for granitic-related deposits such as tin, porphyry, copper, or molybdenum deposits tend to have high salinities, common daughter minerals, and low CO_2 contents (Roedder 1984). This would seem to disfavor a magmatic origin for the fluid in the gold deposits reported here.

4.6.2 Stable Isotopic Data

This section summarizes the available carbon, oxygen, and hydrogen isotopic data from the deposits in eastern Hebei province and discusses their implications for the nature and source of the ore-forming fluids. The sulfur isotopic data from these deposits were discussed in Chapter 4.2.3, and will not be repeated here. The use of stable isotopes for mineral-pair geothermometry was attempted only for the Niuxinshan deposit and the Wangtoushan deposit (Sanjia district), and the results are discussed in Chapters 3.1.7 and 3.2.7.

Carbon Isotopes

A number of studies have attempted to use the carbon isotopic composition of carbonates from mesothermal gold deposits in order to constrain the source of ore-bearing fluids (Kerrich 1987, 1989; Golding et al. 1989; Peters and Golding 1989). Analyses of the carbon and oxygen isotopic composition of vein carbonates from the Niuxinshan and Sanjia districts are shown in Fig. 4.13 together with the fields of possible carbon reservoirs (mantle, metamorphic rocks, and marine limestones) taken from the literature. Also shown for comparison are carbonate compositions from the Archean gold deposits of Canada (Kerrich 1989) and Australia (Golding et al. 1989).

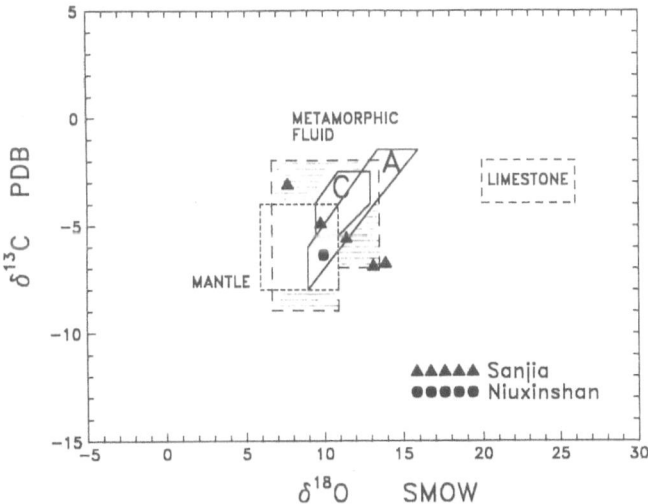

Fig. 4.13. Carbon and oxygen isotopic composition of vein carbonates from the Niuxinshan and Sanjia gold deposits. Also shown for comparison are fields of carbonate composition from Archean gold deposits in Australia (*A*) and Canada (*C*). (Kerrich 1989). Isotopic composition of possible C-O reservoirs are taken from Faure (1986), Valley (1986), and Kerrich (1989)

The number of samples analysed from the Chinese deposits is small (data are given in Tables 3.4 and 3.8), yet all show a restricted range of $\delta^{13}C$ values between -3 and $-6.8‰$, and a wider range of $\delta^{18}O$ values of $+7.7$ to $+13.9‰$. These isotopic compositions of carbonate from the Chinese gold deposits are in the same range as those reported from gold deposits of the Canadian shield and Western Australia. There are two problems with interpreting these data to constrain the source of the ore-bearing fluids. The first is that carbonate minerals in the Chinese gold deposits formed later than the sulfide ore minerals (and gold), and therefore it is not certain that the carbon isotopic data are pertinent to the ore-forming stage of hydrothermal activity. Secondly, as discussed in some detail by Kerrich (1989), values of around $-5‰$ for $\delta^{13}C$ in carbonates are not diagnostic of any single carbon reservoir. The data are consistent with derivation from mantle CO_2 (Golding et al. 1989), magmatic CO_2 from a granitic magma (Burrows and Spooner 1987), or from thoroughly mixed average "crustal carbon" (Ohmoto and Rye 1979). Moreover, the carbon isotopic composition of carbonates is a function of the pH and oxygen fugacity of the fluid from which they formed, because of the strong isotopic fractionation between reduced and oxidized carbon compounds (Ohmoto and Rye 1979). Therefore, it must be concluded that the carbon isotopic data are of little use to constrain the source of ore-bearing fluids.

Oxygen and Hydrogen Isotopes

A compilation of all available oxygen and hydrogen isotopic data from vein minerals in the gold deposits of eastern Hebei province is given in Table 4.7. The hydrogen data represent analyses of fluid released from fluid inclusions in quartz. The table also shows the calculated oxygen isotopic composition of the hydrothermal fluid from each deposit based on a model temperature for mineralization (from fluid inclusion data) and quartz-water fractionation factors of Matsuhisa et al. (1979).

The isotopic composition of quartz is similar for all the deposits, ranging between $+10$ and $+14‰$ relative to SMOW. These values, and the estimated $\delta^{18}O$ values of about $+4$ to $+8‰$ for the hydrothermal fluids, are typical for Archean gold deposits worldwide, as summarized by Groves and Foster (1991), and they are somewhat lower and less variable than those of vein quartz from Phanerozoic mesothermal lode gold deposits as reviewed by Nesbitt (1991). The oxygen isotopic composition of vein quartz in the Chinese deposits overlaps with the reported isotopic composition of quartz from the Archean metamorphic country rocks (i.e., $+13$ to $+15‰$ for quartz from metamorphic iron formations of the Qianxi Group, Qian et al. 1985; and $+11.6‰$ for amphibolite from the Jinchangyu district, Sun et al. 1989). This suggests that the hydrothermal fluids were in isotopic equilibrium with the metamorphic country rocks. It is important in this light to note that there is no difference in the oxygen isotopic composition

Table 4.7. Oxygen and hydrogen isotopic data from quartz veins in gold deposits of eastern Hebei province

Deposit	Sample	Mineral	$\delta^{18}O$ (SMOW) Quartz	δD (SMOW) Fluid inclusion	$\delta^{18}O$ (SMOW) Fluid	Reference[a]
Niuxinshan	6111	Quartz	10.6		3.7	1 TUM
	6020	Quartz	10.9		4.0	1 TUM
	6555	Quartz	11.7		4.8	1 TUM
	6181	Quartz	11.6		4.7	1 TUM
	6111	K-feldspar	8.7			1 TUM
	6020	K-feldspar	9.2			1 TUM
	6555	K-feldspar	11.2			1 TUM
	6181	K-feldspar	10.0			1 TUM
	21	Quartz	13.0	−85.5	6.1	1 MMI
	50	Quartz	12.1	−133.4	5.2	1 MMI
Huajian	CE89059	Quartz	13.3	−64	6.4	1 MMI
Sanjia	64	Quartz	14.4	−85.5	7.5	1 MMI
	CE89020	Quartz	12.5	−81	5.6	1 MMI
	91-1	Quartz	12.8	−48.9	5.9	1 MMI
Wangtoushan	6309	Quartz	10.6		3.7	1 TUM
	6271	Quartz	11.4		4.5	1 TUM
	6273	Quartz	10.2		3.3	1 TUM
	6309	K-feldspar	11.3			1 TUM
	6271	K-feldspar	9.9			1 TUM
	6273	K-feldspar	11.2			1 TUM
Jinchangyu	54-2	Quartz	11.4	−50	4.5	2
	65-3	Quartz	11.2	−73.9	4.3	2
	69-1	Quartz	13.5	−67.7	6.6	2
	70-5	Quartz	11.3	−86.5	4.4	2
	CE89150	Quartz	10.9	−75	4.0	1 MMI
	–	Quartz	10.9			3
	–	Quartz	10.8			3
Yuerya	74-1	Quartz	10.3	−69.4	3.4	2
	Q3	Quartz	13.7	−88.6	6.8	1 MMI
	Q4	Quartz	12.6	−88.6	5.7	1 MMI

[a] References: 1 TUM, this study, analyses by Technical University of Munich; 1 MMI, this study, analyses by Ministry of Metallurgical Industry; 2, Yu and Jia (1989); 3, Yu et al. (1989).

of quartz from the amphibolite-hosted deposits (Niuxinshan, Sanjia, Jinchangyu) and of quartz from the Yuerya deposit, which is hosted by Proterozoic carbonate rocks and granite. This suggests a uniformity of fluid composition which disregards local lithology, i.e., a deep and extensive circulation of fluids on a subregional scale.

The genetic interpretation of this range of $\delta^{18}O$ values is difficult because fluids of both a magmatic and metamorphic source would be consistent with these isotopic values (Taylor 1979). Meteoric water is generally much

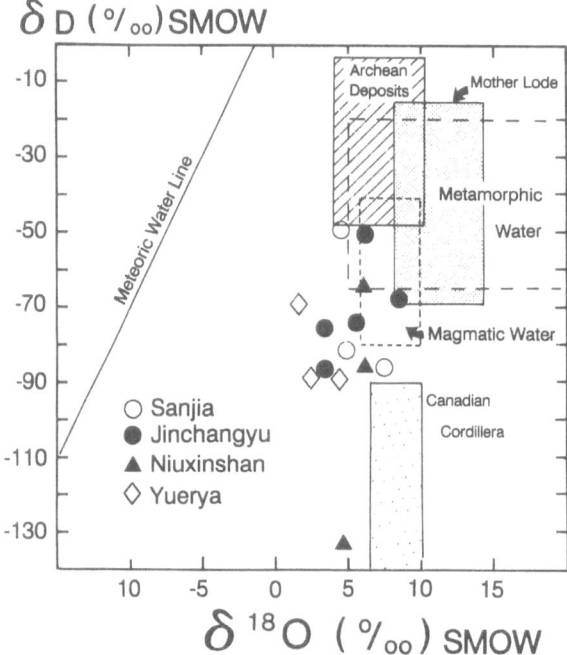

Fig. 4.14. Hydrogen and oxygen isotopic composition of ore fluid from selected gold deposits of eastern Hebei province. Data sources are given in Table 4.7. Also shown for comparison are fields of ore fluid composition from other mesothermal gold deposits (Archean, Mother Lode, Canadian Cordillera) after Nesbitt and Muehlenbachs (1989). Isotopic composition of magmatic and metamorphic fluids are taken from Taylor (1979)

lower in ^{18}O, unless it had thoroughly exchanged oxygen with crustal rocks. Further constraints on fluid source can be made by combined oxygen and hydrogen isotopic analyses. Figure 4.14 shows the limited data available relating to ore fluid from the Chinese gold deposits. The hydrogen values represent measured δD of fluid released from inclusions in vein quartz, and oxygen isotopic compositions of ore fluid were calculated from the $δ^{18}O$ values of quartz as given in Table 4.7. The position of the meteoric water line and fields of magmatic and metamorphic fluids are from Taylor (1979). Also shown for comparison with the Chinese deposits are the inferred compositions of ore fluids reported from other mesothermal gold deposits by Nesbitt and Muehlenbachs (1989).

The hydrogen and oxygen isotopic compositions of ore fluids from all of the Chinese deposits are quite similar, and they plot far to the right of the meteoric water line near the fields of magmatic and metamorphic fluids. A metamorphic fluid can be ruled out because the quartz veins postdate any known regional metamorphism in the area by more than one billion years, and furthermore, the petrographic evidence in the rocks indicates that the youngest metamorphic event was retrograde, which would consume and not release fluid. Yu and Jia (1989) interpreted their oxygen and hydrogen isotopic data from the Jinchangyu deposit as indicating a mixed magmatic-meteoric fluid. The hypothesis of magmatic water may be questioned because the ore veins clearly postdate the emplacement of granites in the mining districts, and because there is no granite in the immediate vicinity of some deposits (notably Jinchangyu). On the other hand, late magmatic dikes are coeval with quartz veins in most districts, and therefore a magmatic origin for the fluids cannot be dismissed. The hypothesis of thoroughly equilibrated "crustal water" of meteoric origin is consistent with the uniformity of fluid inclusion and stable isotopic compositions (S, C, H, O) from vein minerals among several districts, and it would also fit the lead isotopic evidence that the ore lead was leached from a high-grade metamorphic basement. The oxygen isotopic evidence given above for equilibration of the ore-fluids with the metamorphic country rocks also supports the possibility that the fluids were "disguised" meteoric waters.

We conclude that the stable isotopic evidence is permissive of more than one explanation for the source of ore fluids. We consider that the best explanation of the isotopic, fluid inclusion, and geologic evidence is given by the hypothesis that the ore fluids were meteoric waters whose isotopic composition was thoroughly equilibrated with the metamorphic basement rocks.

4.6.3 Gold Transport and Precipitation

The evidence discussed above suggests that the ore-bearing fluids in the gold deposits of eastern Hebei province were at a temperature of at least 300 °C (fluid inclusion homogenisation), and consisted of water with about 10 mol% CO_2 and a total salinity of about 1 M NaCl equivalent (4–8 wt%). The coexistence of K-feldspar and sericite in the wall-rock selvages of most veins suggests near-neutral pH. The sulfur isotopic evidence, lack of sulfate minerals, and the detection of coexisting CH_4 and CO_2 in inclusion fluids suggest that the fluids were reduced. With these fluid characteristics as constraints, the possible mechanisms of gold transport and deposition can be discussed.

Seward (1984, 1989, 1991) has summarized the hydrothermal geochemistry of gold, and an important conclusion is that there is still much uncertainty about the speciation of gold in solution, thermodynamic data being absent for many potential gold complexes, particularly at high temperature. It is commonly assumed that the solubility of gold in natural hydrothermal fluids depends on the stability of one or both of the two gold complexes shown by the following equations:

$$Au^\circ + H_2S + HS^- = Au(HS)_2^- + 0.5H_2, \tag{1}$$

$$Au^\circ + H^+ + 2Cl^- = AuCl_2^- + 0.5H_2. \tag{2}$$

The bisulfide complex (1) is dominant under reducing conditions at near-neutral pH, and at relatively low temperature; and there is little doubt that this species is important for gold transport and deposition in epithermal deposits and active geothermal systems (Berger and Henley 1989; Seward 1989). At the higher temperatures of ore formation in mesothermal gold deposits, however, both the bisulfide and the dichloride complexes may be important. Many studies of Archean greenstone-belt gold deposits have emphasized the role of the bisulfide complex (Groves and Foster 1991) because of the following lines of evidence:

1. the low salinity, neutral pH, and reducing conditions of the ore fluids,
2. the association of gold with sulfidization of high Fe/Fe + Mg rocks, and
3. the very low enrichments of Cu, Pb, and Zn, which are typically carried in solution as metal-chloride complexes.

The fluid composition inferred for the Chinese gold deposits is similar to the composition of ore fluids in the Archean examples cited by Groves and Foster (1991). Furthermore, as discussed in Chapter 4.3.3, a correlation of gold with sulfidization of iron-rich rocks can be locally observed in the Chinese deposits. This suggests that a significant amount of gold in the Chinese examples was also transported in the ore fluid as sulfide complexes. On the other hand, the Chinese ore fluids are slightly more saline than many of the Archean examples cited by Groves and Foster (1991), and many of the Chinese ores contain large amounts of base metal sulfides. Therefore, the role of chloride complexes for the transport of gold in these deposits may also be important. Sang and Ho (1987) stated that gold was transported as chloride complexes in the Mesozoic gold districts of northeastern China which they investigated.

We conclude that both sulfide and chloride complexes were probably important for the transport of gold in the deposits of eastern Hebei province. Possible causes of gold precipitation can be inferred from the stability of these complexes, but critical data about the ore fluids are lacking (i.e., pH, oxygen fugacity, concentration of total sulfur, etc.) to evaluate their real importance. Both complexes would be destabilized by mixing of ore fluid with cooler, more dilute fluids. Seward (1991) pointed out that boiling (or effervescence) of ore fluids would also cause the deposition of gold from both sulfide and chloride complexes. However, the limited fluid inclusion data available from the Chinese deposits suggest that the ore fluids were not boiling at the time of trapping.

The ubiquity of intense wall-rock alteration around the ore veins in the Chinese deposits suggests that fluid-rock interaction may be important in gold precipitation. Breakdown of the bisulfide complex due to pyritization of wall rocks (removal of reduced sulfur from the fluid) was discussed in Chapter 4.3.3 as one probable mechanism of gold precipitation. The sericitization of feldspars is a very common type of wallrock alteration in the Chinese deposits. This reaction consumes hydrogen ions and therefore raises the fluid pH, which would have the effect of destabilizing the gold-chloride complex (Seward 1984). Redox reactions will also affect the solubility of gold, with opposite affects on chloride complexes (stable under oxidizing conditions) and on sulfide complexes (stable under reducing conditions), but the redox state in the veins and alteration zones around the Chinese deposits has not been studied and the importance of this factor cannot be evaluated at present.

5 Towards a Metallogenetic Model

Table 5.1 summarizes the main characteristics of the gold deposits in eastern Hebei province documented in this study. The most important geologic and geochemical constraints on a successful metallogenetic model are listed here together with chapter references to where the evidence was introduced in the text:

1. There is some evidence for a relict, subeconomic, syn-metamorphic mineralization in the Jinchangyu and Banbishan deposits (Chap. 3.4.5, 3.5.5), but the main-stage ore veins in all districts studied are discordant and post-metamorphic. On a regional scale, the deposits are related to major NE–SW-striking crustal-scale faults and lineaments. The ore-bearing veins occur in subsidiary compressional (reverse and oblique-slip) faults formed or reactivated in the Early Yanshanian Orogeny (Jurassic). Dilation and vein filling may have been caused by a rotation of the regional stress field (Chap. 4.5.2).

2. The ore paragenesis in all districts studied is simple and uniform. The paragenetic sequence begins with barren vein quartz and pyrite. Most of the gold occurs later, together with galena, sphalerite and chalcopyrite in fractures of early formed pyrite. Accessory ore minerals include tetrahedrite, tetradymite, native bismuth, molybdenite, tellurides, and scheelite. The fineness of the gold is fairly uniform at 600 to 800. The bulk Au/Ag ratio in the ores is generally less than one (Chap. 3).

3. Most gold deposits and occurrences are found within exposures of high-grade metabasic rocks of the Archean Qianxi Group (Chap. 1.3). Of secondary importance are Early Proterozoic metasedimentary host rocks. Granites are rarely host to ore, with the important exception of the Yuerya deposit.

4. The gold concentrations determined in unmineralized host rocks of the Qianxi Group fall in the range of average crustal values (1–5 ppb) in most samples (Chap. 4.2.1). These values may not represent the primary gold concentrations, but the available evidence suggests that the protoliths were not enriched in gold. Nevertheless, the Qianxi Group rocks are considered to be the ultimate source of gold in the deposits (Chap. 4.2.5).

5. The gold deposits and occurrences are closely related both spatially and temporally to Late Mesozoic Yanshanian granites and associated igneous dikes. The gold-bearing quartz veins generally cut the Yanshanian

163

Table 5.1. Summary of the main characteristics of the investigated gold deposits in eastern Hebei province

Deposit	Type of ore body	Ore minerals	Alteration minerals	Host rocks	Age of mineralization	Stable isotopes (vein minerals)	P-T and fluid conditions
Niuxinshan	Quartz veins in amphibolite, Disseminations in granites	Pyrite, chalcopyrite, galena sphalerite, gold, covellite tetradymite, bismuth scheelite	Sericite, quartz, carbonate, pyrite, chlorite, minor K-feldspar, fluorite	Archean amphibolites, migmatite gneisses, magnetite amphibolites; Yanshan granites and dikes	190 ± 2 Ma (Rb-Sr, sericite) 174 ± 25 Ma (Rb-Sr, K-feldspar) 176 ± 3 Ma (Ar-Ar, quartz) 188 Ma (K-Ar, sericite)	Carbonate: $\delta^{13}C$ = -6.3, $\delta^{18}O$ = $+9.7$ Pyrite: $\delta^{34}S$ = $+4.9$ to $+6.0$ Quartz: $\delta^{18}O$ = $+10.6$ to $+11.7$ Fluid inclusions: δD = -85.5	CO_2-H_2O fluid, 10 mol% CO_2 6-9 wt. % NaCl equiv. Minimum temperature = 270–315°C Minimum pressure = 1.5–2 kbar
Sanjia	Quartz veins in amphibolite	Pyrite, chalcopyrite, galena, sphalerite, gold, tetrahedrite, bismuth, pyrrhotite	Sericite, quartz, carbonate, pyrite, chlorite	Archean amphibolites, magnetite amphibolites; Yanshan dikes	168 ± 3 Ma (Ar-Ar, quartz)	Carbonate: $\delta^{13}C$ = -5.5, $\delta^{18}O$ = $+11.4$ Pyrite: $\delta^{34}S$ = -1.0 to $+3.6$ Quartz: $\delta^{18}O$ = $+12.5$ to $+14.4$ Fluid inclusions: δD = -81 to -85	CO_2-H_2O fluid, No microthermometric data available
Xinglonggou	Quartz veins in amphibolite	Pyrite, chalcopyrite, galena, sphalerite, gold, covellite, tetrabedrite	Sericite, quartz, carbonate, pyrite, chlorite	Archean amphibolites, magnetite amphibolites; Yanshan dikes	No data available	Carbonate: $\delta^{13}C$ = -6.8 Carbonate: $\delta^{18}O$ = 13.2	CO_2-H_2O fluid, No microthermometric data available

Deposit	Occurrence	Ore minerals	Alteration minerals	Isotope data	Age	Fluid characteristics
Wangtoushan	Quartz veins in amphibolite, Disseminations in granite	Pyrite, chalcopyrite, galena, sphalerite, gold, tetrahedrite, molybdenite, scheelite	Sericite, quartz, carbonate, pyrite, chlorite, minor fluorite, K-feldspar	184 ± 2 Ma (Rb-Sr, sericite) 179 ± 25 Ma (Rb-Sr, K-feldspar)	Archean amphibolites, migmatite gneisses, magnetite amphibolites; Yanshan granites and dikes	Carbonate: $\delta^{13}C = -5.8$ Carbonate: $\delta^{18}O = +11.9$ Quartz: $\delta^{18}O = +10.2$ to $+11.4$	CO_2-H_2O fluid, 10 mol% CO_2 6–8 wt. % NaCl equiv. Minimum temperature = 190–340°C Minimum pressure = 2.5 kbar
Yuerya	Quartz veins, disseminations in granite	Pyrite, chalcopyrite, galena, sphalerite, gold, pyrrhotite, molybdenite, tetrahedrite, chalcocite, calaverite	Albite, sericite, chlorite, quartz, pyrite	$<165 \pm 5$ Ma (K-Ar, granite) 200 Ma (K-Ar, sericite)	Yanshan granite and dikes; Proterozoic limestone	Pyrite: $\delta^{34}S = +1.5$ to $+4.5$ Quartz: $\delta^{18}O = +10.3$ to $+13.7$ Fluid inclusions: $\delta D = -69$ to -88	CO_2-H_2O fluid, no salinity data Minimum temperature = 260–390°C No pressure estimate
Jinchangyu	Quartz veins in amphibolite	Pyrite, chalcopyrite, galena, sphalerite, gold, calaverite, argentite, molybdenite, pyrrhotite, chalcocite, hessite	Sericite, carbonate, quartz, pyrite, chlorite	192 Ma (K-Ar, sericite) 169 Ma (K-Ar, sericite) 155 Ma (K-Ar, sericite) 133 Ma (Pb-Pb, galena)	Archean amphibolites, migmatic gneisses, magnetite amphibolites; Yanshan dikes	Pyrite: $\delta^{34}S = -6$ to $+2.5$ Quartz: $\delta^{18}O = +10.8$ to $+13.5$ Fluid inclusions: $\delta D = -68$ to -85	CO_2-H_2O fluid, no salinity data Minimum temperature = 256–370°C No pressure estimate

Table 5.1 (*continued*)

Deposit	Type of ore body	Ore minerals	Alteration minerals	Host rocks	Age of mineralization	Stable isotopes (vein minerals)	P-T and fluid conditions
Banbishan	Quartz veins in leptite, Disseminations in shear zones	Pyrite, chalcopyrite, galena, sphalerite, gold, arsenopyrite, pyrrhotite, wolframite, scheelite, antimonite	Sericite, K-feldspar, quartz, carbonate, chlorite, pyrite	Proterozoic leptite, metadiorite; Yanshan dikes	No data available	No data available	No data available

intrusions and are coeval with some of the dikes. This suggests that mineralization took place at a late stage of the Yanshanian magmatism (Chap. 3).

6. The radiometric dating of mineralization confirms the Late Mesozoic age inferred from field relations. In the gold deposits of eastern Hebei province for which data are available (Niuxinshan, Huajian, Wangtoushan, Jinchangyu, Yuerya) the ages of hydrothermal minerals range from 150 to 190 Ma (Chap. 4.1.2).

7. The role of the Yanshanian granites in mineralization is not entirely clear. Gold is associated with both magnetite-series hornblende-biotite granodiorites and with ilmenite-series leucocratic biotite granites. The granites apparently did not constitute the source of gold or other ore-forming elements in the deposits studied (Chap. 4.4).

8. The physical conditions of mineralization were similar in all districts studied. Fluid inclusion studies suggest that the mineralization occurred at a minimum of about 300–350 °C and 2–4 kbar pressure. The ore fluid had low to moderate salinity and about 10 mol% CO_2 (Chap. 4.6.1).

9. Wall rock alteration is strongly developed to a distance of a few centimeters to some meters from the veins. The alteration involved the breakdown of feldspars, amphiboles, pyroxene, and biotite, and the formation of chlorite, pyrite, sericite, and hydrothermal quartz. Carbonate is also common in the alteration zones, but was generally later than the main phase of alteration and mineralization (Chap. 3).

10. The sulfidization of magnetite to pyrite in magnetite-rich host rocks is an important type of wall rock alteration which locally caused gold precipitation. However, magnetite-rich lenses are thin and discontinuous in the gold districts of Niuxinshan, Sanjia, and Jinchangyu, and do not constitute a major host to ore (Chap. 4.3).

11. The common lead isotopic composition of galena from ore veins suggests that the lead was derived from Early Precambrian U-depleted rocks (Chap. 4.2.3). Carbon and sulfur isotopic data permit the interpretation that the the Precambrian basement rocks were also the source of sulfur and carbon in the fluids; however, the data do not rule out a magmatic origin for these elements (Chap. 4.2.3, 4.6.2).

12. The oxygen and hydrogen isotopic compositions of ore fluid, calculated from analyses of vein quartz and of water from fluid inclusions, are similar in all deposits studied despite local differences in wallrock composition (Chap. 4.6.2). The isotopic data suggest that the hydrothermal fluid was either of magmatic origin, or was meteoric water thoroughly equilibrated with the Precambrian metamorphic rocks. The latter interpretation better fits the geologic evidence (Chap. 4.6.2).

5.1 Previous Models

The gold deposits of eastern Hebei province have been studied by Chinese geologists for many years and several hypotheses of their origin have been proposed. Following Zhu (1989), the models can be divided essentially into two rival types as follows:

1. Metamorphic-Hydrothermal Models
The metamorphic hydrothermal models consider that the ultimate source of gold is in the mafic and ultramafic Archean rocks of the lower crust. The agent for leaching and transporting the gold and other ore-forming elements is considered to be low salinity CO_2-H_2O fluid developed by devolatilization reactions during prograde metamorphism of the lower crust. The fluids are channeled by crustal-scale fault zones, along which the gold deposits eventually form. Recent proponents of the metamorphic model include Metamorphic models are also favored by some authors for the Archean greenstone-belt gold deposits in Australia (Groves et al. 1987, 1988) and Canada (Card et al. 1989; Colvine 1989). In these cases, there is a general concordancy of the age of mineralization and the age of regional metamorphism. However, the main problem with a metamorphic model for the eastern Hebei deposits is that the age of mineralization is Mesozoic, whereas the last regional metamorphic event in the area is Late Proterozoic. There is, in fact, some evidence for a minor syn-metamorphic gold mineralization in the Jinchangyu and Banbishan deposits (Chap. 3.4.5, 3.5.5) but this early mineralization is of no economic significance.
A second problem with the metamorphic model is that it fails to explain the strong spatial association of gold deposits and Mesozoic granites. Some proponents of the metamorphic-hydrothermal model recognize this problem and suggest that the granites remobilized an earlier, synmetamorphic mineralization (Zhu 1989; Wang and Sun 1989).

2. Magmatic Hydrothermal Models
The magmatic hydrothermal models are presently most favored for the gold deposits in northeastern China (Sang and Ho 1987; Wang and Cheng 1988; M.Z. Yang 1988; Feng and Yang 1989; Liu 1989; Zhou 1989). These models propose that the Yanshanian granites were the agent of gold transport and concentration from the source area to the site of intrusion. The primary source of gold and other ore-forming elements is not tightly constrained by the models. Many authors consider that the granites are derived by anatexis of lower crustal rocks, and that the gold originates in that source (Sang and Ho 1987; M.Z. Yang 1988; Zhu 1989). The source of hydrothermal fluids in the various models may be either magmatic volatiles or meteoric water which is brought into circulation by the heat of the intrusions.

The magmatic models have the important advantage that they are consistent with the mineralization ages in the gold deposits of eastern Hebei province, and they explain the spatial relationships between gold deposits and Yanshanian granites. However, the models are not without problems. First, the question must be raised of why only some of the Yanshanian granites are associated with gold deposits whereas most are not. Second, the style of the supposed granite-related gold mineralization is quite different from that of unequivocal granite-related deposits of Cu, Mo, Sn, and W. In the latter cases, mineralization is clearly centered on granite intrusions or their contacts with the country rocks, and the granites themselves are important hosts of ore. In the gold deposits of eastern Hebei province, the granites are rarely mineralized and the locus of gold-bearing veins is by no means centered around the intrusions.

5.2 The Preferred Model for Eastern Hebei Province

The authors believe that the geologic and geochemical constraints listed above are best explained by a "magmatic-hydrothermal" model, to use the terminology of Zhu (1989). The deposits were directly caused by processes related to the Yanshanian Orogeny, but we also stress the role of the Archean basement rocks as a source of gold and other ore elements. The preferred model is based on three indisputable facts:

First, the mineralization is epigenetic; second, the gold deposits occur almost exclusively within Early Precambrian metamorphic terranes; and third, the main stage of mineralization in all the gold districts studied is Mesozoic in age.

The source of gold in the deposits is considered to be the Early Precambrian metamorphic basement, which consists mostly of an Archean high-grade gneiss terrane (not a greenstone belt) containing mantle-derived metabasalts, iron formations, and TTG-type orthogneisses metamorphosed to upper amphibolite and granulite facies conditions. The present gold concentration in the Archean rocks is generally in the range of 1–5 ppb. Whether this range reflects the original gold contents of the protoliths is unknown. The granulite-facies metamorphism at about 2500 Ma (Fuping Orogeny) and a subsequent medium-grade metamorphic event at about 1800 Ma (Zhongtiao Orogeny) may well have mobilized gold from the protoliths, but evidence of syn-metamorphic mineralization is very rare, the only known examples being at Jinchangyu and, possibly Banbishan (see Chap. 4.1.1). The mobilization of gold by metamorphic fluids and its concentration in higher-level shear zones, which is thought to be important for the formation of greenstone-belt gold deposits worldwide (Groves et al. 1987; Colvine 1989), was either unimportant in the Chinese examples or evidence for it has been removed by erosion. It must be kept in mind that the crustal level exposed in the Chinese Archean terrane spans the granulite-amphibolite facies transition, whereas in the Western Australian and Canadian greenstone terranes, the greenschist facies is exposed.

The most significant event for the formation of the gold deposits in northeastern China, indeed, for metallogeny in eastern Asia as a whole, was the Mesozoic Yanshanian Orogeny. In eastern Hebei province, the Yanshanian Orogeny mainly caused widespread compressional faulting, often along preexisting structures, and the intrusion of numerous relatively small granitic plutons and multifarious dikes. The ultimate cause of the Yanshanian magmatism is considered to be the subduction of oceanic lithosphere beneath the Sino-Korean Platform (Takahashi 1983; Ishihara 1984; Wu 1985). Shallow subduction of relatively hot lithosphere can be inferred from the plate reconstructions, and this explains the wide belt of Yanshanian magmatism extending far inland from the continental margin (Takahashi 1983). Kerrich and Wyman (1990) have suggested that mesothermal gold deposits of Archean and Phanerozoic age are preferentially located in convergent margin settings, and are ultimately caused by subduction-related crustal underplating and attendant processes. This tectonic setting is valid for the deposits in northeastern China as well.

According to the few isotopic data available (low initial $^{87}Sr/^{86}Sr$ ratios, nonradiogenic common Pb), the Yanshanian granites in eastern Hebei province apparently formed by partial melting of the lower crust in response to this broad thermal disturbance (M.Z. Yang 1988). Their intrusion was facilitated by pre-existing major fault zones, so that they now occur along regional lineaments and at their mutual intersections. The intrusion of many individual granites in a narrow time interval represents a very significant convective transfer of heat into the upper crust. This heat is considered to be far more important for the formation of the gold deposits than the magmas themselves, since the gold deposits are associated with granites of quite different bulk compositions, and there is no compelling evidence that any of the ore constituents were derived from the granites.

Equally important for the metallogenetic model as the granitic magmatism was the creation and/or reactivation of major compressional fault zones during the Yanshanian Orogeny. These brittle and brittle-ductile faults and fracture zones acted as zones of weakness which facilitated granite intrusion and subsequent dike formation. Additionally, the fault zones provided a focus for fluid circulation which was driven by temperature gradients around the cooling plutons, perhaps aided in the brittle regime by a seismic-pumping mechanism (Eisenlohr et al. 1989). Gold mineralization formed in secondary splays of the major faults, in dilatent zones which formed either as a result of local fault geometry (intersections, jogs, competency contrasts) or due to changing regional stress patterns. The association of ore veins with granites may be partly due to the structural effect of competency contrast.

The source of ore fluid can be only poorly constrained. Hydrogen and oxygen isotopic data from vein quartz are permissive of magmatic, metamorphic, or surface water thoroughly exchanged with crustal rocks. Metamorphic fluids can be ruled out because prograde metamorphism had long ceased before mineralization. Unadulterated meteoric water can also be ruled out, but a choice between magmatic and "equilibrated" surface water, or an admixture of the two, is difficult. The low salinity of the ore fluids argues against a magmatic fluid since fluid inclusions from magmatogenic porphyry deposits are characteristically very saline (Roedder 1984; but see Burrows and Spooner 1987). Furthermore, the mineralization postdates crystallization of the main granites, being coeval only with volumetrically minor dikes. We therefore favor the interpretation that the ore fluids were derived from surface waters whose isotopic composition was thoroughly equilibrated with the basement rocks. Extensive interaction of ore fluids with the metamorphic basement rocks on a regional scale is attested to by the following lines of evidence:

1. The uniform composition of ore fluids from widely-separated gold deposits, inferred from the similarity of wallrock alteration and fluid inclusion data, and from isotopic measurements of ore and gangue minerals (C, S, Pb, O, H), could best be achieved by large-scale equilibration of fluid with the basement rocks.
2. The isotopic composition of lead in vein sulfides from all of the deposits studied (including those hosted in granite) suggest that the lead was derived from leaching of the basement rocks. The carbon and sulfur isotopic data are not diagnostic of a single source, but the data are consistent with a derivation of these elements from the basement rocks as well.

The gold in solution was most likely transported at the conditions of mineralization (300–350°C, 2–4 kbar) as both chloride and sulfide complexes. Sulfide complexes are suggested by the clear correlation of gold with sulfidization of wallrocks (see below), and the presence of chloride complexes is inferred from the relatively high temperatures, moderate salinities (equivalent to about 1 M NaCl) of the fluids, and the association of abundant base metals in the ores (see also Sang and Ho 1987).

Precipitation of gold from either sulfide or chloride complexes would be effected by cooling of the hydrothermal fluid or by phase separation (boiling or CO_2 effervescence) caused by local pressure release. The association of gold with other ore minerals near wallrock selvages and rarely within the vein quartz suggests that chemical effects of wallrock interaction must have played the most important role in precipitation. However, the lack of quantitative estimates of the pH, oxidation potential, activity of sulfide and chloride ligands in the fluid, etc., and the fact that some of these parameters have opposite effects on the stability of chloride and sulfide complexes of gold, preclude a complete description of how gold was precipitated in these ores. The almost exclusive occurrence of gold with sulfide minerals in the ores suggests that the precipitation of gold relates to the sulfides. The explanation for this may be that sulfide formation lowered the activity of reduced sulfur in solution and thereby caused the instability of the bisulfide gold complex. This is certainly the case for gold enrichment in pyritized lenses of magnetite quartzite, and it is a common mechanism suggested for gold precipitation in other mesothermal deposits (Neall 1987; Groves and Foster 1991). A possible precipitation mechanism involving the chloride gold complex could be the sericitization of feldspars, which is a very common type of wall rock alteration. The stability of the gold-chloride complex is highest at low pH. Sericitization would have the effect of neutralizing acidic fluids, and this could cause the breakdown of the complex.

The significance of the metallogenetic model to exploration for gold deposits in this setting is that deposits are most likely to be found in areas where the three elements: major fault zones, Archean crust, and Yanshanian granites, occur together. The country rocks around granites in zones of high fault density, especially those which occur at or near intersections of regional fault zones, are favourable targets for exploration. On a local scale, gold mineralization is most likely to be found in sulfide-rich veins in mafic host rocks with well-developed alteration zones.

5.3 Open Questions

Several questions about the genesis of gold deposits in eastern Hebei province remain only partially answered, partly due to lack of data and partly due to the geologic and geochemical complexity of the problem. Some of the more important points which need further study are suggested here.

The reader will have noticed that the amount of information presented is not uniform from all deposits. In particular, studies on fluid inclusions, age dating, and stable isotopic work are lacking or have been done only in a reconnaisance fashion in several deposits. The most severe lack of information at the time of writing concerns the deposits located in the Early Proterozoic metamorphic rocks (Banbishan and related districts), and these should perhaps be a priority for further research.

It is still uncertain to what extent pre-Yanshanian mineralization is important to the distribution or grade of the gold deposits. The evidence is overwhelming that the main stage – and in many deposits the only detectable stage – of mineralization is Mesozoic, but there is also clear evidence for pre-Mesozoic mineralization in the Jinchangyu and Banbishan deposits. It may be significant that the largest deposit in eastern Hebei province, Jinchangyu, is the one which shows the clearest evidence for an earlier stage of mineralization.

An important point which needs further work is the significance of iron-rich host rocks in localizing ore formation. Field observations indicate that lenses of metamorphosed iron formation which occur in the gold mining districts are strongly sulfidized where crossed by ore veins, and the available analyses show a significant correlation between sulfidization and gold concentration. The importance of iron-rich host rocks has been abundantly demonstrated in Western Australian gold deposits (Neall 1987; Groves and Foster 1991). On the other hand, the major iron mines operating in Archean BIF in eastern Hebei province do not encounter gold mineralization to any significant extent (Shen et al. 1989).

The true relationship between Yanshanian granites and gold metallogeny is still poorly defined. In particular, it is impossible at present to understand why certain granites are associated with gold and others are not. A major hindrance to this understanding is the lack of data on the precise age, composition, petrographic nature, and intrusive style of the dozens of Yanshanian granites in eastern Hebei province and surroundings. As a single example, the lack of U and Th data makes it impossible to discuss the importance of internal heat production as a factor in promoting hydrothermal circulation around certain granites.

The igneous dikes associated with granites have been only cursorily studied. Many districts display a common intrusive sequence beginning with granitic dikes followed by dioritic dikes and ending with lamprophyre dikes, with mineralization generally coeval with the diorite dikes. The significance of this sequence has not been explored. Note, too, that the possible role of lamprophyres in gold metallogeny of Western Australia and Canada has recently been stressed by Rock et al. (1989) and Wyman and Kerrich (1989).

More detailed mineralogical-geochemical studies aimed at determining the critical parameters of ore fluid composition are needed to constrain models of gold transport and deposition, and to better interpret the sulfur and carbon isotope data. More fluid inclusion studies, including Raman spectroscopic data instead of bulk fluid analyses, and detailed studies of alteration geochemistry would be particularly helpful.

6 Comparison with Other Archean-Hosted Gold Provinces

The amount of research and published information on mesothermal vein-type gold deposits in general, and in Archean terranes especially, has increased dramatically in recent years. The intention of this chapter is not to review the nature of mesothermal gold deposits in general, but simply to point out some important similarities and differences between the Archean-hosted gold deposits in eastern Hebei province and the well-known Archean-hosted gold deposits of the Canadian shield, Western Australia, and Southern Africa. For information on the latter areas we depend heavily on previous reviews by Hutchinson and Vokes (1987), Groves et al. (1988), Card et al. (1989), Foster (1989), Vearncombe et al. (1989), and Groves and Foster (1991). Before discussing individual features of the Chinese deposits, it is worth stating that there are far more similarities than differences between these deposits and other Archean-hosted mesothermal gold deposits. In fact, Nesbitt (1991) and Kerrich and Wyman (1990) have stressed that mesothermal gold deposits of all ages share a great many features and should perhaps be treated as a single class of ore deposit.

Host Rocks

The gold deposits in eastern Hebei province are hosted almost exclusively in Early Precambrian metamorphosed supracrustal rocks, and by far the most important examples are hosted in multiply deformed, high-grade Archean metabasic rocks (migmatitic amphibolites and mafic granulites). Orthogneisses are common in the Archean exposures but they rarely host important gold mineralization. It must be stressed that the Archean terrane in northeastern China represents a high-grade gneiss terrane in the sense of Windley (1984) and it is not just an intensely metamorphosed greenstone belt. In this sense the Chinese example is unusual, since most of the Archean-hosted gold deposits in other cratons are located in greenstone belts and the adjacent gneiss terranes are unimportant (Groves et al. 1987; Card et al. 1989). A further important point is that, although BIF horizons are common in the Archean terrane in northeastern China, they do not host gold deposits. In contrast, BIF is a very important host to ore in Western Australia and Zimbabwe (Groves et al. 1988; Foster 1989).

The close spatial association of gold deposits with granite intrusions is an important feature of eastern Hebei province. Similar associations have been noted for the Canadian shield by Card et al. (1989), and for the Norseman-

Wiluna belt in Western Australia by Groves et al. (1988). The significance of the granite association is controversial in both cases, and as noted in Chapter 5.3, the role of granites in gold metallogeny of eastern Hebei province is also not completely clear.

Structural Control

The gold deposits in eastern Hebei province are strictly confined to faults and/or fracture zones of brittle and brittle-ductile nature. On a district scale the deposits are located near major, regional faults of a compressional nature. These features are common to a great many other gold provinces in Archean terranes around the world (Vearncombe et al. 1989; Groves and Foster 1991), and point to the importance of regional-scale faults in focusing hydrothermal circulation.

Ore Fluids

Like the structural setting, the fluid inclusion characteristics form one of the unifying aspects of mesothermal gold deposits worldwide. The data from deposits in eastern Hebei province indicate that ore fluids were of low to moderate salinity (4–8 wt% NaCl equivalent), contained about 10 mol% CO_2, and were trapped in vein minerals at temperatures of at least 300–350 °C. With the exception that many Archean deposits have somewhat lower ranges of salinity (often less than 4 wt% NaCl equiv.), these fluid characteristics are identical with those from many Archean examples (Ho et al. 1990; Groves and Foster 1991) and Phanerozoic examples (Nesbitt 1991). The source of ore fluids is also fully as ambiguous in the Chinese examples as in many others (Groves et al. 1988)!

Metal Association

The element association in gold ores from eastern Hebei province (Au, Ag, Cu, Pb, Zn, Bi, W, ±Mo) is somewhat different from that in many Archean-hosted gold deposits. The latter have been called gold-only deposits because of the lack of base metal enrichment and the high Au/Ag ratios of about 10 in bulk ores (Groves and Foster 1991; Fyfe and Kerrich 1984). In the Chinese examples reported here, in contrast, base metals are commonly present at the percent level in the ores, and the bulk Au/Ag ratios are invariably less than unity. Both of these features could be explained by a higher salinity of the ore fluids in the Chinese case, since the transport of base metals is primarily as chloride complexes, and silver is incorporated in galena.

Tectonic Setting

Recent integrated studies of the tectonic setting of Archean gold deposits in the Norseman-Wiluna Belt suggests an assocation of the deposits with accretionary continental margins (Barley et al. 1989). Similar ideas have been expressed for the Superior Province in Canada by Card et al. (1989). Kerrich and Wyman (1990) suggested that mesothermal lode gold deposits of all ages share a common geodynamic setting of continental margin accretion. According to these authors, the unifying features of mesothermal gold deposits including late- to postmetamorphic timing, association with crustal-scale faults, low salinity C-O-H fluids, and dominantly volcanic host rocks can be explained by processes following accretion by subduction of oceanic crust.

The gold deposits in eastern Hebei province, and probably elsewhere in northeastern China as well, formed in an unambiguous convergent margin setting. The deposits formed during the Mesozoic Yanshanian Orogeny, which involved eastward subduction of oceanic crust beneath the Sino-Korean platform margin. The gold deposits therefore were the result of the magmatic and tectonic activity associated with this Mesozoic reactivation of the Archean craton.

Appendix 1. Average composition of selected Yanshanian granitic plutons from eastern Hebei province

Pluton	Sanyihe	Dushan	Niuxinshan	Wangtoushan	Xiaoyingzi	Gaojiadian	Jiajiashan	Laochengling
SiO_2	73.93	70.07	74.25	75.69	68.79	61.48	73.93	76.66
TiO_2	0.12	0.24	0.07	0.08	0.38	1.18	0.10	0.07
Al_2O_3	13.42	14.68	13.12	12.91	14.26	15.91	13.82	12.62
Fe_2O_3	1.06	1.99	1.15	1.00	3.01	5.58	1.59	1.15
MgO	0.18	0.57	0.13	0.12	1.01	1.61	0.21	0.20
MnO	0.22	0.03	0.17	0.01	0.08	0.08	0.05	0.05
CaO	0.54	1.54	0.58	0.96	1.83	3.53	0.49	0.37
Na_2O	4.54	4.32	3.32	3.06	4.37	4.66	4.68	4.27
K_2O	4.25	4.37	5.09	4.96	4.23	3.52	4.05	3.77
P_2O_5	0.02	0.14	0.05	0.01	0.16	0.84	0.11	0.02
LOI	0.36	0.40	1.17	0.73	0.65			
Total	98.51	98.05	98.59	99.11	98.23	98.39	99.00	99.18
Rb	144	71	292	173	60	157	150	
Sr	93	721	29	195	554	508	70	
Ba	272	2113	230	651	1394	1823	127	
Cr	22	46	4	6	17	9	8	
Zr	99	43	85	96	228	18	88	
Zn	44	48	90	37	40	60	32	
Y	14	6	29	9	16	26	4	
Nb	26	13	64	20	12	27	43	
La	16.7	30.0	8.7	7.5		53.3	11.7	
Ce	35.8	51.9	19.3	17.5		131.0	32.4	
Nd	14.7	16.0	9.9	8.0		40.7	7.7	
Sm	2.9	2.2	2.9	1.4		7.1	1.4	
Eu	0.4	0.5	0.2	0.2		1.8	0.3	
Gd	2.3	1.6	2.4	0.9		5.4	1.0	
Tb	0.7	0.3				0.8	0.3	
Dy	1.9	0.9	2.5	0.8		4.1	1.0	
Yb	2.7	0.4				2.3	0.7	
Lu	0.2	0.1	0.3	0.2		0.4	0.2	
n	7	9	12	7	13	6	2	1

n, number of samples.
Total Fe as Fe_2O_3.

Appendix 1 (*continued*)

Pluton	Louwenyu	Madi	Maoshan	Qianfengshuiling	Qingshankou	Shitaizi	Wangpingshi	Yuerya
SiO_2	75.35	74.15	74.46	69.76	73.57	73.52	73.36	73.97
TiO_2	0.10	0.03	0.13	0.24	0.20	0.18	0.18	0.12
Al_2O_3	13.18	14.61	13.29	15.45	13.67	12.82	14.07	13.09
Fe_2O_3	1.34	0.73	1.72	2.44	1.64	2.57	1.61	2.02
MgO	0.28	0.11	0.28	0.82	0.38	0.43	0.31	0.25
MnO	0.18	0.12	0.12	0.09	0.07	0.03	0.09	0.06
CaO	0.56	0.29	1.32	1.62	0.59	1.03	0.67	0.80
Na_2O	4.18	5.44	3.97	4.66	4.34	3.70	4.22	3.51
K_2O	4.11	3.63	3.86	3.62	4.39	4.57	4.48	4.33
P_2O_5	0.08	0.07	0.14	0.18	0.57	0.18	0.12	0.06
LOI								
Total	99.37	99.17	99.28	98.87	99.40	99.02	99.11	98.21
Rb	275	1590	333		155		233	138
Sr	51	4	239		113		138	158
Ba	196	10	551		524		495	403
Cr	7	8	7		8		6	8
Zr	73	73	97		44		73	166
Zn	46	59	52		53		35	256
Y	29	10	28		11		12	22
Nb	46	166	58		29		35	24
La		8.8	20.3		46.0		21.3	
Ce		32.4	48.2		109.4		46.0	
Nd		14.0	19.0		47.1		12.0	
Sm		5.9	4.8		9.0		2.2	
Eu		0.1	0.6		2.2		0.3	
Gd		4.8	4.6		7.3		1.8	
Tb		0.8	0.8		1.1		0.4	
Dy		2.7	4.7		5.7		1.6	
Yb		0.5	3.6		2.6		2.2	
Lu		0.1	0.7		0.4		0.5	
n	2	4	4	3	2	2	3	12

n, number of samples.
Total Fe as Fe_2O_3.

Appendix 2. Partial analyses including gold of rocks from the Qianxi Group in eastern Hebei province

Sample	CE97	CE98	CE99	CE100	CE101	CE102	CE103	Q25	CE208
Rock type	Websterite	Hornblendite	Websterite	Websterite	Websterite	Websterite	Amphibolite	Amphibolite	Granulite
SiO_2	42.06	42.98	52.64	51.62	53.18	43.88	43.96	48.40	49.64
TiO_2	0.23	0.22	0.14	0.19	0.19	0.17	1.48	0.65	0.42
Al_2O_3	6.88	5.06	3.62	4.82	3.59	5.43	6.09	9.40	4.05
Fe_2O_3	15.32	11.58	9.43	9.70	9.90	11.44	17.36	9.80	11.88
MgO	25.52	26.86	24.84	18.93	24.56	27.34	16.39	14.65	17.34
MnO	0.12	0.07	0.12	0.10	0.10	0.11	0.18	0.15	0.18
CaO	5.38	5.88	8.01	10.38	6.76	4.63	9.25	12.93	12.57
Na_2O	0.34	0.30	0.30	1.28	0.54	0.64	1.12	1.22	0.99
K_2O	0.18	0.08	0.08	0.28	0.18	0.36	0.08	1.08	0.25
P_2O_5	0.09	0.05	0.05	0.01	0.01	0.18	0.16	0.21	0.02
S(tot)	–	–	–	–	–	–	–	–	–
Au (ppb)	6	<1	10	1	1	1	1	2	1
Pd (ppb)	45	28	67	18	14	32	4	11	10
Ag	–	–	–	–	–	–	–	–	–
Cu	–	–	–	–	–	–	–	–	–
Zn	–	–	–	–	–	–	–	–	–
Pb	–	–	–	–	–	–	–	–	–
W	–	–	–	–	–	–	–	–	–
Cr	–	–	–	–	–	–	–	–	–
Co	–	–	–	–	–	–	–	–	–
Ni	–	–	–	–	–	–	–	–	–

Appendix 2 (*continued*)

Sample Rock type	CE209 Hornblendite[a]	CE106 Hornblendite[a]	CE93 Amphibolite[a]	CE96 Hornblendite[a]	CE104 Amphibolite[a]	CE105 Hornblendite[a]	CE95 Hornblendite[b]	CE201 Hornblendite[b]
SiO_2	47.98	43.78	43.06	44.18	41.56	43.76	44.12	49.80
TiO_2	0.50	1.99	1.85	1.77	2.17	2.02	1.89	1.46
Al_2O_3	7.81	6.20	5.26	6.10	6.85	7.59	4.68	6.29
Fe_2O_3	12.63	18.05	19.80	17.64	19.04	17.93	19.05	15.32
MgO	17.10	16.83	16.67	15.40	15.29	12.97	15.40	17.96
MnO	0.15	0.12	0.13	0.10	0.34	0.46	0.12	0.14
CaO	11.74	7.59	8.75	9.75	10.75	12.12	8.63	6.70
Na_2O	1.27	0.80	1.22	0.88	1.04	1.46	1.04	1.32
K_2O	0.50	0.28	0.28	0.64	0.46	0.64	0.18	0.12
P_2O_5	0.09	0.23	0.46	0.27	0.34	0.46	0.25	0.04
S(tot)	–	–	–	–	–	–	–	–
Au (ppb)	1	1	1	1	2	2	98	32
Pd (ppb)	2	5	3	2	5	5	2	2
Ag	–	–	–	–	–	–	–	–
Cu	–	–	–	–	–	–	–	–
Zn	–	–	–	–	–	–	–	–
Pb	–	–	–	–	–	–	–	–
W	–	–	–	–	–	–	–	–
Cr	–	–	–	–	–	–	–	–
Co	–	–	–	–	–	–	–	–
Ni	–	–	–	–	–	–	–	–

Appendix 2 (*continued*)

Sample Rock type	CE202 Pyroxenite	CE203 Hornblendite[b]	CE205 Hornblendite[a]	CE204 Amphibolite	YQ26 Granulite	YQ28 Leptite	YQ31 Leptite	YQ56 Amphibolite
SiO_2	52.12	40.90	42.10	50.52	50.96	48.84	54.84	48.84
TiO_2	0.78	0.74	1.66	0.66	0.72	0.92	0.54	1.25
Al_2O_3	8.44	11.59	10.52	12.17	14.82	14.20	16.39	12.84
Fe_2O_3	9.80	28.09	24.10	11.20	11.73	13.81	8.34	14.24
MgO	13.17	7.90	9.32	9.28	8.10	6.65	5.45	6.95
MnO	0.12	0.36	0.08	0.09	0.19	0.25	0.13	0.19
CaO	11.99	8.26	10.03	9.86	10.60	12.40	8.80	11.95
Na_2O	2.04	1.48	1.56	2.90	3.38	3.04	4.22	4.80
K_2O	1.40	0.94	0.58	1.24	1.32	1.00	1.68	0.64
P_2O_5	0.09	0.04	0.10	0.18	0.23	0.13	0.22	0.14
S(tot)	–	–	–	–	–	–	–	–
Au (ppb)	2	16	2	1	1	2	1	4
Pd (ppb)	2	3	2	5	12	7	8	22
Ag	–	–	–	–	–	–	–	–
Cu	–	–	–	–	–	–	–	–
Zn	–	–	–	–	–	–	–	–
Pb	–	–	–	–	–	–	–	–
W	–	–	–	–	–	–	–	–
Cr	–	–	–	–	–	–	–	–
Co	–	–	–	–	–	–	–	–
Ni	–	–	–	–	–	–	–	–

Appendix 2 (*continued*)

Sample Rock type	YQ90 Amphibolite[a]	YG124 Amphibolite	YQ8 Pyroxenite	YQ14 Plagiogneiss	YQ19 Plagiogneiss	YQ22 Plagiogneiss	YQ25 Amphibolite	YQ26 Amphibolite
SiO_2	48.22	49.00	–	–	–	–	–	–
TiO_2	1.36	0.94	0.82	0.65	0.58	0.50	0.47	0.55
Al_2O_3	13.20	13.39	–	–	–	–	–	–
Fe_2O_3	16.26	12.42	16.20	6.41	9.39	13.40	12.80	11.60
MgO	6.50	8.70	21.56	5.31	7.63	7.30	6.96	8.12
MnO	0.29	0.17	0.15	0.08	0.09	0.18	0.14	0.13
CaO	11.15	12.74	7.00	6.02	5.88	11.89	9.93	8.54
Na_2O	3.00	2.72	1.05	4.03	3.41	3.28	3.49	3.45
K_2O	0.88	0.48	0.30	2.77	1.57	0.67	0.72	0.81
P_2O_5	0.09	0.09	0.14	0.27	0.32	0.07	0.14	0.23
S(tot)	–	–	–	–	–	–	–	–
Au (ppb)	1	1	50	53	3	4	2	1
Pd (ppb)	4	10	2	2	3	2	3	12
Ag	–	–	<0.5	<0.5	<0.5	<0.5	<0.5	<0.5
Cu	–	–	200	140	150	200	150	83
Zn	–	–	120	91	110	96	94	120
Pb	–	–	10	18	18	10	16	14
W	–	–	–	–	–	–	–	–
Cr	–	–	1200	140	490	320	250	360
Co	–	–	90	23	36	54	40	39
Ni	–	–	800	59	160	120	88	76

Appendix 2 (*continued*)

Sample Rock type	YQ28 Leptite	YQ30 Felsic gneiss	YQ31 Leptite	YQ38 Pyroxenite	YQ41 Gneiss	YQ45 Amphibolite	YQ48 Orthogneiss	YQ56 Amphibolite
SiO_2	–	–	–	–	–	–	–	–
TiO_2	0.73	0.75	0.45	0.72	0.37	0.40	0.22	0.87
Al_2O_3	–	–	–	–	–	–	–	–
Fe_2O_3	14.70	14.10	9.33	14.20	12.30	12.60	35.40	16.80
MgO	6.30	7.30	5.64	4.81	10.45	10.11	2.49	6.63
MnO	0.17	0.17	0.10	0.14	0.13	0.12	0.07	0.17
CaO	10.91	10.77	7.00	12.03	11.19	10.49	4.20	9.23
Na_2O	3.38	3.44	4.52	3.29	2.75	2.87	1.61	3.14
K_2O	0.63	0.58	1.05	0.43	1.01	1.06	0.22	0.53
P_2O_5	0.14	0.11	0.21	0.09	0.05	0.05	0.16	0.16
S(tot)	–	–	–	–	–	–	–	–
Au (ppb)	2	1	1	6	2	3	5	4
Pd (ppb)	7	7	8	3	3	2	3	22
Ag	<0.5	<0.5	<0.5	<0.5	<0.5	<0.5	<0.5	<0.5
Cu	140	320	85	140	90	130	270	63
Zn	110	100	90	89	83	76	73	130
Pb	10	14	24	12	18	14	8	14
W	–	–	–	–	–	–	–	–
Cr	250	150	220	140	270	480	30	170
Co	46	60	30	35	51	50	11	55
Ni	72	81	52	40	130	140	16	83

Appendix 2 (*continued*)

Sample Rock type	YQ72 Orthogneiss	YQ73 Gneiss	YQ78 Websterite	YQ82 Amphibolite	YQ87 Plagiogneiss	YQ88 Pyroxenite	YQ95 Gneiss	YQ98 Amphibolite
SiO_2	–	–	–	–	–	–	–	–
TiO_2	0.68	0.48	0.70	0.60	0.53	0.62	2.00	0.57
Al_2O_3	–	–	–	–	–	–	–	–
Fe_2O_3	16.90	12.60	27.20	14.70	12.90	14.70	17.70	14.80
MgO	6.47	5.47	4.15	7.13	8.29	11.11	4.48	8.29
MnO	0.19	0.12	0.18	0.17	0.14	0.17	0.22	0.14
CaO	10.21	7.84	6.72	9.37	8.40	15.39	9.09	8.12
Na_2O	2.34	3.91	2.34	2.70	2.96	1.73	2.98	3.16
K_2O	0.33	0.65	0.36	0.84	0.95	1.02	0.75	0.67
P_2O_5	0.09	0.11	0.21	0.09	0.07	1.81	0.23	0.07
S(tot)	–	–	–	–	–	–	–	–
Au (ppb)	3	10	1	4	11	98	4	1
Pd (ppb)	2	9	2	2	2	3	2	2
Ag	<0.5	<0.5	<0.5	<0.5	<0.5	<0.5	<0.5	<0.5
Cu	190	88	84	65	80	120	240	260
Zn	120	75	140	110	88	170	130	110
Pb	10	10	12	16	12	14	12	18
W	–	–	–	–	–	–	–	–
Cr	430	300	310	280	390	330	160	260
Co	60	33	32	55	48	45	60	56
Ni	120	86	28	95	94	130	39	120

Appendix 2 (*continued*)

Sample Rock type	YQ101 Pyroxenite	YQ108 Hornblendite	YQ111 Pyroxenite	YQ123 Plagiogneiss	YQ124 Amphibolite	YQ130 Amphibolite	YQ133 Amphibolite
SiO_2	–	–	–	–	–	–	–
TiO_2	1.27	0.73	0.52	0.47	0.43	3.34	0.04
Al_2O_3	–	–	–	–	–	–	–
Fe_2O_3	17.50	15.70	12.40	12.40	13.70	17.00	53.20
MgO	5.64	12.10	7.30	10.94	11.61	6.13	3.65
MnO	0.21	0.15	0.12	0.13	0.13	0.15	0.15
CaO	8.54	9.79	8.54	7.28	10.91	6.16	5.04
Na_2O	3.18	2.58	4.00	2.78	1.77	3.76	0.08
K_2O	0.76	0.90	0.46	0.45	0.42	0.84	0.04
P_2O_5	0.14	0.48	0.16	0.11	0.05	0.25	0.18
S(tot)	–	–	–	–	–	–	–
Au (ppb)	2	1	1	1	1	1	12
Pd (ppb)	2	2	2	2	10	2	2
Ag	<0.5	<0.5	<0.5	<0.5	<0.5	<0.5	<0.5
Cu	240	130	290	180	130	200	190
Zn	110	160	120	88	83	130	39
Pb	16	10	16	12	12	12	8
W	–	–	–	–	–	–	–
Cr	160	420	190	310	220	10	20
Co	65	57	42	52	56	53	5
Ni	95	110	71	120	130	59	1

Major element oxides in wt%, trace elements in ppm or ppb as indicated, total Fe as Fe_2O_3.
Au and Pd analyses by INAA following fire assay (Au) or nickel sulfide (Pd) concentration.
[a] Sample taken from BIF zone.
[b] Sample taken from fracture zone.

186

References

Barley ME, Eisenlohr B, Groves DI, Perring CS, Vearncombe JR (1989) Late Archean convergent margin tectonics and gold mineralization: a new look at the Norseman-Wiluna Belt, Western Australia. Geology 17: 826–829

Berger BR, Henley RW (1989) Advances in the understanding of epithermal gold-silver deposits, with special reference to the western United States. Econ Geol Monogr 6: 405–423

Boyle RW (1979) The geochemistry of gold and its deposits. Geol Surv Can Bull 280

Brown GC, Plant J, Lee MK (1979) Geochemical and geophysical evidence on the geothermal potential of Caledonian granites in Britain. Nature 280: 129–131

Brown PE (1989) FLINCOR: A microcomputer program for the reduction and investigation of fluid-inclusion data. Am Mineral 74: 1390–1393

Brown PE, Lamb WM (1989) P-V-T properties of fluids in the system $H_2O \pm CO_2 \pm$ NaCl: New graphical presentations and implications for fluid inclusion studies. Geochim Cosmochim Acta 53: 1209–1221

Brugman GE, Arndt T, Hofmann AW and Tobschall HJ (1987) Noble metal abundances in komatiite suites from Alexo, Ontario, and Gorgona Island, Colombia. Geochim Cosmochim Acta 51: 2159–2169

Burnham CW, Ohmoto H (1980) Late-stage processes of felsic magmatism. Mining Geol Spec Issue 8: 1–11

Burrows DR, Spooner ETC (1987) Generation of a magmatic H_2O-CO_2 fluid enriched in Au and W within an Archean sodic granodiorite stock, Mink Lake, northwestern Ontario. Econ Geol 82: 971–986

Card KD, Poulsen KH, Robert F (1989) The Archean Superior Province of the Canadian shield and its lode gold deposits. Econ Geol Monogr 6: 19–36

Chen GD (1989) Tectonics of China. International Academic Publishers, Pergamon Press, Oxford

Chinese Academy of Geological Sciences (1979) Tectonic Map of the People's Republic of China, scale 1:4000000. Cartographic Publ House, Beijing

Clayton N, Goldsmith JR, Mayeda TK (1989) Oxygen isotope fractionation in quartz, albite, anorthite, and calcite. Geochim Cosmochim Acta 53: 725–733

Colvine AC (1989) An empirical model for the formation of Archean gold deposits: products of final cratonization of the Superior Province, Canada. Econ Geol Monogr 6: 37–53

Crocket JH (1991) Distribution of gold in the Earth's crust. In: Foster RP (ed) Gold metallogeny and exploration. Blackie, London, pp 1–36

Dickinson WR (1979) Plate tectonic evolution of the north Pacific rim. In: Uyeda S, Murphy RQ, Kobayashi K (eds) Geodynamics of the Western Pacific. Advances in Earth Planet Sci 6. Center for Academic Publ, Tokyo, pp 1–20

Doe BR, Stacey JS (1974) The application of lead isotopes to the problems of ore genesis and ore prospect evaluation: A review. Econ Geol 69: 757–776

Doe BR, Zartman RE (1979) Plumbotectonics, the Phanerozoic. In: Barnes HL (ed) Geochemistry of hydrothermal ore deposits. Wiley, New York, pp 22–70

Eisenlohr BN, Groves D, Partington GA (1989) Crustal-scale shear zones and their significance to Archaean gold mineralization in Western Australia. Mineral Depos 24: 1–8

187

Ernst WG, Cao R, Jiang J (1988) Reconnaisance study of Preambrian metamorphic rocks, northeastern Sino-Korean shield, People's Republic of China. Geol Soc Am Bull 100: 692–701

European Community (1990) Gold deposits prospection and exploration technology (China). Final Rep Proj No CI1*-0127-D(BA). Commission of the European Communities, Directorate-General for Science, Research and Technology DGXII, Brussels (unpubl Tech Rep)

Faure G (1986) Principles of isotope geology, 2nd edn. Wiley, New York

Feng SZ, Yang TD (1989) On magmatic hydrothermal metallogenesis of gold in Archean greenstone belts in the North China Platform. In: Guan GY, Zhu FS (eds) Proc Int Symp on Gold geology and exploration, 26–30 June 1989, Shenyang. Northeast University of Technology Publ House, Shenyang, pp 63–69

Foster RP (ed) (1984) Gold '82: The geology, geochemistry and genesis of gold deposits. AA Balkema, Amsterdam

Foster RP (1989) Archean gold mineralization in Zimbabwe: implications for metallogenesis and exploration. Econ Geol Monogr 6: 54–70

Foster RP (ed) (1991) Gold metallogeny and exploration. Blackie, London

Foster RP, Gilligan JM (1987) Archean iron-formation and gold mineralization in Zimbabwe. In: Appel UPW, Laberge GL (eds) Precambrian iron-formations. Theophrastus, Athens, pp 635–674

Fyfe WS, Kerrich R (1984) Gold: Natural concentration processes. In: Foster RP (ed) Gold '82: the geology, geochemistry and genesis of gold deposits. AA Balkema, Amsterdam, pp 99–128

Fyon JA, Troop DG, Marmont S, Macdonald AJ (1989) Introduction of gold into Archean crust, Superior Province, Ontario—coupling between mantle-initiated magmatism and lower crustal thermal maturation. Econ Geol Monogr 6: 479–490

Gao DY (1986) Geological features and metallogenic mechanism of Jinchangyu gold deposit in Hebei province. Changchun Insititute of Gold Research, Ministry of Metallurgical Industry, Gold and Silver Anthology 5: 140–148 (in Chinese)

Gao ZL, Lin EW (1987) The study of the whole-rock gold abundance of the Jinchangyu gold mine. J Changchun Coll Geol 17: 65–72 (in Chinese)

Grant JA (1986) The isocon diagram, a simple solution to Gresens' equation for metasomatic alteration. Econ Geol 81: 1976–1982

Golding SD, McNaughton NJ, Barley ME, Groves DI, Ho SE, Rock NMS, Turner JV (1989) Archean carbon and oxygen reservoirs: their significance for fluid sources and circulation paths for Archean mesothermal gold deposits of the Norseman-Wiluna Belt, Western Australia. Econ Geol Monogr 6: 376–388

Groves DI, Foster RP (1991) Archean lode gold deposits. In: Foster RP (ed) Gold metallogeny and exploration. Blackie, London, pp 63–103

Groves DI, Phillips N, Ho SE, Houstoun SM, Standing CA (1987) Craton-scale distribution of Archean greenstone gold deposits: predictive capacity of the metamorphic model. Econ Geol 82: 2045–2058

Groves DI, Ho SE, McNaughton NL, Mueller AG, Perring CS, Rock NMS, Skwarnecki MS (1988) Genetic models for Archean lode gold deposits in Western Australia. In: Ho SE, Groves DI (eds) Advances in understanding Precambrian gold deposits, vol II. The Geology Department and University Extension Publ 12, University of Western Australia, Nedlands, pp 1–22

Groves DI, Barley ME, Ho SE (1989) Nature, genesis, and tectonic setting of mesothermal gold mineralization in the Yilgarn Block, Western Australia. Econ Geol Monogr 6: 71–85

Guan GY (1988) Secondary source beds and Precambrian lode gold deposits in the northern China platform. In: Goode ADT, Smyth EL, Birch WD, Bosma LI (eds) Abstracts and proceedings, gold '88. May 1988, Melbourne. Geol Soc Aust Abstr Ser 23, pp 593–595

Guan GY, Zhu FS (eds) (1989) Proc Int Symp on Gold Geology and Exploration, 26–30 June 1989, Shenyang. Northeast University of Technology Publ House, Shenyang

Guan GY, Jin CZ, Wu XH (1989) Monosource-polygenetic model of gold deposits in Liaoning Province and its adjacent areas. In: Guan GY, Zhu FS (eds) Proc Int Symp on Gold geology and exploration, 26–30 June 1989, Shenyang. Northeast University of Technology Publ House, Shenyang, pp 72–76

Guilbert JM, Park CF Jr (1986) The geology of ore deposits. WH Freeman, New York

Guo WK (1982) On granitoids relevant to metallogeny. Reg Geol China 2: 15–30 (in Chinese)

Guo WK (1987) Metallogenic map of endogenic ore deposits of China with guide, scale 1:4 000 000. Cartographic Publ House, Beijing (in Chinese and English)

Hamlyn PR, Keays RR, Cameron W, Crawford AJ, Waldron HM (1985) Precious metals in magnesian low-Ti lavas: implications for metallogenesis and sulfur saturation in primary magmas. Geochim Cosmochim Acta 49: 1797–1811

Hao ZP (1989) Geochemical feature of the Dongfengshan gold deposit. In: Guan GY, Zhu FS (eds) Proc Int Symp on Gold geology and exploration. 26–30 June 1989, Shenyang. Northeast University of Technology Publ House, Shenyang, pp 330–333

Harris LB (1987) A tectonic framework for the Western Australian Shield and its significance to gold mineralization: a personal view. In: Ho SE, Groves DI (eds) Recent advances in understanding Precambrian gold deposits. Geology Department and University Extension Publ 11. The University of Western Australia, Nedlands, pp 307–320

Ho SE, Groves DI, Phillips GN (1990) Fluid inclusions in quartz veins associated with Archean gold mineralization: clues to ore fluids and ore depositional conditions and significance to exploration. In: Herbert HK, Ho SE (eds) Stable isotopes and fluid processes in mineralization. Geology Department and University Extension Publ 23. The University of Western Australia, Nedlands, pp 35–50

Hu AG (1989) Geological setting and the genesis of Jiapigou gold deposit. In: Guan GY, Zhu FS (eds) Proc Int Symp on Gold geology and exploration. 26–30 June 1989, Shenyang. Northeast University of Technology Publ House, Shenyang, pp 346–352

Huang DY (1986) Fundamental metallization pattern of gold-silver deposits in the northwestern part of Shandong Peninsula. Geol Prospect 12: 10–15 (in Chinese)

Huang JQ (1945) On major tectonic forms of China. Natl Geol Surv China Geol Mem Ser A 20 (in Chinese)

Huang X, Bi Z, DePaulo DJ (1986) Sm-Nd isotope study of Early Archean rocks, Qianan, Hebei province, China. Geochim Cosmochim Acta 50: 625–635

Hutchinson CS, Taylor D (1978) Metallogenesis in SE Asia. J Geol Soc Lond 135: 407–428

Hutchinson RW, Vokes FM (1987) Introduction: special issue on Precambrian gold deposits. Econ Geol 82: 1991–1992

Ikonnikov AB (1975) Mineral resources of China. Geol Soc Am Microform Publ 2, Boulder, CO

Ishihara S (1981) Granitoid series and mineralization. Econ Geol 75th Anniversary Vol, pp 418–484

Ishihara S (1984) Granitoid series and Mo/W-Sn mineralization in east Asia. Geol Surv Jpn Rep 263: 173–208

Ishihara S, Sato T (1982) Mineral resources of China. Part 3. Granitoids of southern China. Chishitsu News 340: 34–45

Ishihara S, Lee DS, Kim SY (1981) Comparative study of Mesozoic granitoids and related W-Mo mineralization in southern Korea and southwestern Japan. Mining Geol 31: 311–320

Iyama JT, Fonteilles M (1981) Mesozoic granitic rocks of southern Korea reviewed from major constituents and petrography. Mining Geol 31: 281–296

Jahn BM (1990) Origin of granulites: geochemical constriants from Archean granulite facies rocks of the Sino-Korean craton, China. In: Vielzeuf D, Vidal P (eds) Granulites and crustal differentiation, NATO ASI Series. Kluwer, Dordrecht, pp 471–492

Jahn BM (1991) Early Precambrian rocks of China. In: Hall RP, Hughes DJ (eds) Early Precambrian basic magmatism. Blackie, Glasgow, pp 294–315

Jahn BM, Ernst WG (1990) Late Archean Sm-Nd isochron age for mafic-ultramafic supracrustal amphibolites from the northeastern Sino-Korean craton, China. Precambrian Res 46: 295–306

Jahn BM, Zhang ZQ (1984) Archean granulite gneisses from eastern Hebei province, China: rare earth geochemistry and tectonic implications. Contrib Mineral Petrol 85: 224–243

Jahn BM, Auvray B, Cornichet J, Bai YL, Shen QH, Liu DY (1987) 3.5 Ga old amphibolites from eastern Hebei province, China: field occurrence, petrography, Sm-Nd isochron age and REE geochemistry. Precambrian Res 34: 311–346

Keays RR (1984) Archean gold deposits and their source rocks: the upper mantle connection. In: Foster RP (ed) Gold '82: The geology, geochemistry and genesis of gold deposits. AA Balkema, Amsterdam, pp 17–51

Keays RR, Scott RB (1976) Precious metals in ocean-ridge basalts: implications for basalts as source rocks for gold mineralization. Econ Geol 71: 705–719

Keays RR, Ramsay WRH, Groves DI (eds) (1989) The geology of gold deposits: the perspective in 1988. Econ Geol Monogr 6

Kerrich R (1987) The stable isotope geochemistry of Au-Ag vein deposits in metamorphic rocks. Mineral Assoc Canada Short Course Handbook 13, Toronto, pp 287–336

Kerrich R (1989) Archean gold: relation to granulite formation or felsic intrusions? Geology 17: 1011–1015

Kerrich R, Wyman D (1990) Geodynamic setting of mesothermal gold deposits: an association with accretionary tectonic regimes. Geology 18: 882–885

Kim OJ, Lee DS (1983) Summary of igneous activity in South Korea. Geol Soc Am Mem 159: 87–104

Klimetz MP (1983) Speculations on the Mesozoic plate tectonic evolution of eastern China. Tectonics 2: 139–166

Kramers JD, Foster RP (1984) A reappraisal of lead isotope investigations of gold deposits in Zimbabwe. In: Foster RP (ed) Gold '82: the geology, geochemistry and genesis of gold deposits. AA Balkema, Rotterdam, pp 569–582

Lambert IB, Phillips GN, Groves DI (1984) Sulphur isotope compositions and genesis of Archean gold mineralization, Australia and Zimbabwe. In: Foster RP (ed) Gold '82: the geology, geochemistry and genesis of gold deposits. AA Balkema, Rotterdam, pp 373–388

Large RR, Huston DL, McGoldrick PJ, Ruxton PA, McArthur G (1988) Gold distribution and genesis in Australian volcanogenic massive sulfide deposits and their significance for gold transport models. Econ Geol Monogr 6: 520–533

Lee M (1981) Geology and metallic mineralization associated with Mesozoic granitic magmatism in South Korea. Mining Geol 31:235–244

Lhotka PG, Nesbitt BE (1989) Geology of mineralized and gold-bearing iron formation, Contwoyto Lake-Point Lake region, Northwest Territories, Canada. Can J Earth Sci 26: 46–64

Li CF, Liu XS (1986) Geological feature of porphyritic gold deposit in Tuanjiegou. Changchun Institute for Gold Research, Ministry of Metallurgical Industry. Gold Silver Anthol 4: 69–73 (in Chinese)

Li CY, Wang Q, Zhang ZM, Liu XY (1980) A preliminary study of plate tectonics of China. Chin Acad Geol Sci Bull Ser I 2: 11–22

Li CY, Wang Q, Liu XY, Tang YQ (1982) Tectonic Map of Asia, scale 1:8 000 000 with Explanatory Notes. Research Institute of Geology, Chinese Academy of Geological Sciences. Cartographic Publ House, Beijing

Li J (1988) Geological feature and genesis of gold deposits in east Hebei province. Geol Prospect 1: 5–8 (in Chinese)

Liu EW (1985) Research on Pb isotopes in gold deposits in the central part of E. Hebei province. J Changchun Coll Geol 4: 1–9 (in Chinese)

Liu DY, Shen QH, Zhang ZQ, Jahn BM, Auvray B (1990) Archean crustal evolution in China: U-Pb geochronology of the Qianxi complex. Precambrian Res 48: 223–244

Liu LD (1989) A new understanding of magmatic hydrothermal gold deposits. In: Guan GY, Zhu FS (eds) Proc Int Symp on Gold geology and exploration. 26–30 June 1989, Shenyang. Northeast University of Technology Publ House, Shenyang, pp 120–122

Liu JL (1987) Gold deposits in Precambrian BIF. Geol J 1: 58–71 (in Chinese)

Lu GY, Huang JH (1987) New results of Rb-Sr isotopic age for eastern Hebei low grade metamorphic rocks and their geological significance. China Reg Geol 3: 219–224 (in Chinese)

Lu ZX, Fan YX, Sun FY (1989) Structural and magmatic controls over gold mineralization in the Zhaoye gold ore zone, Shandong Province. In: Guan GY, Zhu FS (eds) Proc Int Symp on Gold geology and exploration. 26–30 June 1989, Shenyang. Northeast University of Technology Publ House, Shenyang, pp 135–140

Ma XY, Wu ZW (1981) Early tectonic evolution of China. Precambrian Res 14: 185–202

Maruyama S, Liou JG, Seno T (1989) Mesozoic and Cenozoic evolution of Asia. In: Ben-Avraham Z (ed) The evolution of the Pacific ocean margins. Oxford Monograph on Geology and Geophysics 8, pp 75–99

Matsuhisa Y, Goldsmith JR, Clayton RN (1979) Oxygen isotope fractionation in the system quartz-albite-anorthite-water. Geochim Cosmochim Acta 43: 1131–1140

McElhinny BJ, Embleton BJJ, Ma XH, Zhang ZK (1981) Fragmentation of Asia in the Permian. Nature 293: 212–216

Molnar P, Tapponnier P (1975) Cainozoic tectonics of Asia: the effects of a continental collision. Science 189: 419–426

Neall FB (1987) Sulfidation of iron-rick rocks as a precipitation mechanism for large Archean gold deposits in Western Australia: thermodynamic confirmation. In: Ho SE, Groves DI (eds) Recent advances in understanding Precambrian gold deposits. Department of Geology and University Extension Publ 11. The University of Western Australia, Nedlands, pp 265–270

Nesbitt BE (1991) Phanerozoic gold deposits in tectonically active continental margins. In: Foster RP (ed) Gold metallogeny and exploration. Blackie, London, pp 104–132

Nesbitt BE, Muehlenbachs K (1989) Geology, geochemistry, and genesis of mesothermal lode gold deposits of the Canadian cordillera: evidence for ore formation from evolved meteoric water. Econ Geol Monogr 6: 553–563

Oberthür T, Saager R, Tomschi HP (1990) Geological, mineralogical and geochemical aspects of Archean banded iron formation-hosted gold deposits: some examples from Southern Africa. Mineral Depos 25: 125–135

191

Ohmoto H (1986) Stable isotope geochemistry of ore deposits. In: Valley JW, Taylor HP Jr, O'Neil JR (eds) Stable isotopes in high temperature geological processes. Reviews in mineralogy 16. Mineralogical Society of America, Washington, pp 491–560

Ohmoto H, Rye RO (1979) Isotopes of sulfur and carbon. In: Barnes HL (ed) Geochemistry of hydrothermal ore deposits. Wiley, New York, pp 509–567

O'Neil JR (1986) Theoretical and experimental aspects of isotopic fractionation. In: Valley JW, Taylor HP Jr, O'Neil JR (eds) Stable isotopes in high temperature geological processes. Reviews in mineralogy 16. Mineralogical Society of America, Washington, pp 1–40

Pearce JA, Harris NBW, Tindle AG (1984) Trace element discrimination diagrams for the tectonic interpretation of granitic rocks. J Petrol 25: 956–983

Perring CS, Barley ME, Cassidy KF, Groves DI, McNaughton NF, Rock NMS, Bettenay LF, Golding SE, Hallberg JA (1989) The association of linear orogenic belts, mantle-crustal magmatism, and Archean gold mineralization in the eastern Yilgarn Block of western Australia. Econ Geol Monogr 6: 571–584

Peters SG, Golding SD (1989) Geologic, fluid inclusion, and stable isotope studies of granitoid-hosted gold-bearing quartz veins, Charters Towers, northeastern Australia. Econ Geol Monogr 6: 260–273

Phillips GN, Groves DI (1984) Fluid access and fluid-wall rock interaction in the genesis of the Archean gold-quartz vein deposit at Hunt mine, Kambalda, Western Australia. In: Foster RP (ed) Gold '82: the geology, geochemistry and genesis of gold deposits. AA Balkema, Rotterdam, pp 389–416

Phillips GN, Groves DI, Brown IJ (1987) Source requirements for the Golden Mile, Kalgoorlie: significance to the metamorphic replacement model for Archean gold deposits. Can J Earth Sci 24: 1643–1651

Pidgeon RT (1980) 2480-Ma-old zircons from granulite facies rocks from east Hebei province, north China. Geol Rev 26: 198–207

Ren JS, Jiang CF, Zhang ZK, Qin DY (1987) Geotectonic evolution of China. Science Press, Beijing

Rock NMS, Groves DI, Perring CS, Golding SD (1989) Gold, lamprophyres and porphyries: what does their association mean? Econ Geol Monogr 6: 609–624

Roedder E (1984) Fluid inclusions. Reviews in mineralogy 12. Mineralogical Society of America, Washington

Rösler HJ, Lange H (1972) Geochemical tables. Elsevier, Amsterdam

Rybach L (1976) Radioactive heat production; a physical property determined by the chemistry of rocks. In: Strens RG (ed) The physics and chemistry of rocks. Wiley, New York, pp 309–318

Saager R, Meyer M (1984) Gold distribution in Archean granitoids and supracrustal rocks from southern Africa: a comparison. In: Foster RP (ed) Gold '82: The geology, geochemistry and genesis of gold deposits. AA Balkema, Amsterdam, pp 53–70

Saager R, Oberthür T, Tomschi HD (1987) Geochemistry and mineralogy of banded iron formation-hosted gold mineralization in the Gwanda greenstone belt, Zimbabwe. Econ Geol 84: 197–198

Sang JH, Ho SE (1987) A review of gold deposits in China. In: Ho SE, Groves DI (eds) Recent advances in understanding Precambrian gold deposits. The Geology Department and University Extension Publ 11. The University of Western Australia, Nedlands, pp 307–320

Seward TM (1984) The transport and deposition of gold in hydrothermal systems. In: Foster RP (ed) Gold '82: the geology, geochemistry and genesis of gold deposits. AA Balkema, Amsterdam, pp 165–182

Seward TM (1989) The hydrothermal chemistry of gold and its implications for ore formation: boiling and conductive cooling as examples. Econ Geol Monogr 6: 398–404

Seward TM (1991) The hydrothermal geochemistry of gold. In: Foster RP (ed) Gold metallogeny and exploration. Blackie, London, pp 37–62

Sha RZ (1986) Ore prospecting types of gold deposits in Shanxi Province. Shanxi Metallurgical and geological information, Ministry of Metallurgical Industry (in-house journal) 2: 21–23 (in Chinese)

Shelton KL, So CS, Chang JS (1988) Gold-rich mesothermal vein deposits of the Republic of Korea: geochemical studies of the Jungwon gold area. Econ Geol 83: 1221–1237

Shen BF, Luo H, Li JJ, Peng XL (1989) The types and evolution of Archean granitoid-greenstone terranes in the North China Platform. In: Guan GY, Zhu FS (eds) Proc Int Symp on Gold geology and exploration. 26–30 June 1989, Shenyang. Northeast University of Technology Publ House, Shenyang, pp 198–203

Shenyang Institute of Geology and Mineral Resources (1988a) Contributions to the project of regional metallogenetic condition of main gold deposit types in China 1: Heilongjiang province. Geological Publishing House, Beijing (in Chinese with English summaries)

Shenyang Institute of Geology and Mineral Resources (1988b) Contributions to the project of regional metallogenetic condition of main gold deposit types in China 4: southern Liaoning province. Geological Publ House, Beijing (in Chinese with English summaries)

Shenyang Institute of Geology and Mineral Resources (1989a) Contributions to the project of regional metallogenetic condition of main gold deposit types in China 2: eastern Hebei province. Geological Publ House, Beijing (in Chinese with English summaries)

Shenyang Institute of Geology and Mineral Resources (1989b) Contributions to the project of regional metallogenetic condition of main gold deposit types in China 3: Xiaoqinling area in Henan and Shaanxi provinces. Geological Publ House, Beijing (in Chinese with English summaries)

Shenyang Institute of Geology and Mineral Resources (1989c) Contributions to the project of regional metallogenetic condition of main gold deposit types in China 5: Jiaodong area in Shandong province. Geological Publ House, Beijing (in Chinese with English summaries)

Shenyang Institute of Geology and Mineral Resources (1989d) Contributions to the project of regional metallogenetic condition of main gold deposit types in China 6: southwestern Guizhou province. Geological Publ House, Beijing (in Chinese with English summaries)

Sillitoe RH (1989) Gold deposits in Western Pacific island arcs: the magmatic connection. Econ Geol Monogr 6: 274–291

Sills JD, Wang KY, Yan YH, Windley BF (1987a) The Archean high grade gneiss terrane in eastern Hebei province, NE China: geological framework and conditions of metamorphism. In: Park RG, Tarney J (eds) The evolution of the Lewisian and comparable Precambrian high-grade terranes. Geological Society of London Spec Publ 27, pp 297–305

Sills JD, Wang KY, Yan YH, Windley BF, Zhai MG (1987b) Banded iron formations in the Early Precambrian of NE China. In: Appel PWU, LaBerge GL (eds) Precambrian iron-formations. Theophrastus, Athens, pp 487–511

Stacey JS, Kramers JD (1975) Approximation of terrestrial lead isotope evolution by a two-stage model. Earth Planet Sci Lett 26: 207–221

Stone M, Exley CS (1985) High heat production granites of SW England and their associated mineralization: a review. In: Halls C (ed) High heat production (HHP) granites, hydrothermal circulation, and ore genesis. Institution of Mining and Metallurgy, London, pp 571–593

Sun DZ (ed) (1984) The Early Precambrian geology of eastern Hebei province. Tianjin Science and Technology Press, Tianjin (in Chinese with English summary)

Sun DZ, Lu SN (1985) A subdivision of the Precambrian of China. Precambrian Res 28: 137–162

Sun DZ, Wang WY (1984) Discussion about geochronology. In: Sun DZ (ed) The Early Precambrian geology of the eastern Hebei province. Tianjin Science and Technology Press, Tianjin, pp 24–34 (in Chinese)

Sun DZ, Wu CH (1981) The principal geological and geochemical characteristics of the Archean greenstone-gneiss sequences in North China. Geol Soc Aust Spec Publ 7, pp 121–132

Sun DZ, Wang KY, Wang JL, Yang CL, Zhao FM (1989) Studies on auriferous rock series of Archean in eastern Hebei province. In: Contributions to the project of regional metallogenetic conditions of main gold deposit types in China 2: eastern Hebei province. Geological Publ House, Beijing, pp 49–98 (in Chinese with English summary)

Takahashi M (1983) Space-time distribution of late Mesozoic to Early Cenozoic magmatism in east Asia and its tectonic implications. In: Hashimoto H, Uyeda S (eds) Accretion tectonics in the circum-pacific regions. Terra Scientific, Tokyo, pp 69–88

Tang SL (1986) Discussion of geological feature and ore prospecting direction of gold deposits in Jilin Province. Treatise Collection of Gold Deposit Geology. Guilin Geological Prospecting Company, Ministry of Metallurgical Industry (internal report), pp 17–23 (in Chinese)

Taylor HP Jr (1979) Oxygen and hydrogen isotope relationships in hydrothermal mineral deposits. In: Barnes HL (ed) Geochemistry of hydrothermal ore deposits. Wiley, New York, pp 236–277

Taylor RP, Fryer BJ (1984) Rare earth element lithogeochemistry of granitoid mineral deposits. Can Mining Metallurgy Bull 76: 74–84

Terman M (1984) The last 200 million years in eastern Asia: Yanshanian subduction and post-Yanshanian extension. Geol Surv Jpn Rep 263: 27

Trumbull RB, Satir M, Sun Q, Quo D (1990) Geochemistry, Rb-Sr and oxygen isotopic composition of Yanshanian granitoids in the Anjiaygzi gold district, Inner Mongolia, P.R. China: constraints on the age of mineralization. In: International Mineralogical Association 15th General Meeting Abstracts, Beijing, June 28–July 3, 1990. Chinese Institute of Geology and Mineral Resources Printing House, p 278

Tu GZ (1989) Gold Deposits in China. In: Guan GY, Zhu FS (eds) Proc Int Symp on Gold geology and exploration. 26–30 June 1989, Shenyang. Northeast University of Technology Publ House, Shenyang, pp 2–5

Tu GZ, Wang XZ, Chen XP, Li CY, Zhang GX, Zhao ZH (1984) Geochemistry of strata-controlled ore deposits in China, vol 1. Science Publ House, Beijing (in Chinese)

US Bureau of Mines (1990) Mineral commodity summaries, 1990. US Government Printing Office No 1990-254-368/01089

Uyeda S, Miyashiro A (1974) Plate tectonics and the Japanese Islands: a synthesis. Geol Soc Am Bull 85: 1159–1170

Valley JW (1986) Stable isotope geochemistry of metamorphic rocks. In: Valley JW, Taylor HP Jr, O'Neil JR (eds) Stable isotopes in high temperature geological processes. Reviews in mineralogy 16. Mineralogical Society of America, Washington, pp 445–489

Vearncombe JR, Barley ME, Eisenlohr BN, Groves DI, Houstoun SM, Skwarnecki MS, Grigson MW, Partington GA (1989) Structural controls on mesothermal gold mineralization: examples from the Archean terranes of Southern Africa and Western Australia. Econ Geol Monogr 6: 124–134

Wan TF, Zhu H (1989) The tectonic stress field of the Cretaceous-Early Eocene in China. Acta Geol Sin 63: 14–25 (in Chinese)

Wang AJ (1988) The study and discrimination of the source rocks in gold deposits. Contrib Geol Mineral Resour Res 1: 29–38 (in Chinese)

Wang HZ (ed) (1985) Atlas of the paleogeography of China. Institute of Geology, Chinese Academy of Geological Sciences and Wuhan College of Geology, Cartographic Publ House, Beijing (in Chinese with English summaries)

Wang HZ (1987) Geological feature of the Xiaoqinling gold field and genesis of the deposit. Ore Depos Geol 1: 57–67 (in Chinese)

Wang KY, Sun DZ (1989) Gold mineralization in the eastern part of Hebei province, China. In: Guan GY, Zhu FS (eds) Proc Int Symp on Gold geology and exploration. 26–30 June 1989, Shenyang. Northeast University of Technology Publ House, Shenyang, pp 493–494

Wang KY, Yan YH, Yang RY, Chen YF (1985) REE geochemistry of Early Precambrian charnockites and tonalitic-granodioritic gneisses of the Qianan region, eastern Hebei, north China. Precambrian Res 27: 63–85

Wang KY, Zhang RH, Chen YF (1987) Rb-Sr age of the Shanhaiguan polyphase granitic gneisses. Sci Geol Sin 2: 148–151 (in Chinese)

Wang KY, Windley BF, Sills JD, Yan YH (1990) The Archean gneiss complex in eastern Hebei province, north China: geochemistry and evolution. Precambrian Res 48: 245–265

Wang LK, Zhu WF, Zhang SL, Yang WJ (1983) The evolution of two petrogeno-mineralization series and Sr isotopic data from granites in South China. Mining Geol 33: 295–303

Wang RM, He S, Chen Z, Li P, Dai F (1985) Geochemical evolution and metamorphic development of the Early Precambrian in eastern Hebei, China. Precambrian Res 27: 111–129

Wang XZ, Cheng JP (1988) Major geological characteristics and origin of gold deposits in China. In: Goode ADT, Smyth EL, Birch WD, Bosna LI (eds) Abstracts and proceedings, bicentennial gold '88 May 1988 Melbourne. Geol Soc Aust Abstr Ser 23, pp 408–413

Wang YW (1989) Studies of stable isotopic geochemistry of gold deposits in China. In: Guan GY, Zhu FS (eds) Proc Int Symp on Gold geology and exploration. 26–30 June 1989, Shenyang. Northeast University of Technology Publ House, Shenyang, pp 782–789

Wei J (1989) Discussion of the geological character and genesis of the Yuerya gold deposit, Hebei province, PRC. In: Guan GY, Zhu FS (eds) Proc Int Symp on Gold geology and exploration. 26–30 June 1989, Shenyang. Northeast University of Technology Publ House, Shenyang, p 498

Wiley TJ, Howell DG, Wong FL (1990) Terrane analysis of China and the Pacific rim. Circum-Pacific Council for Energy and Mineral Resources. Earth Science Series 13. The Circum-Pacific Council for Energy and Mineral Resources, Houston

Wilson M (1989) Igneous petrogenesis. Unwin Hyman, London

Windley BF (1984) The evolving continents, 2nd edn. Wiley, Chicester New York

Wu LR (1985) Mesozoic granitoids in East China. In: The crust—the significance of granites-gneisses in the lithosphere. Theophrastus, Athens, pp 201–215

Wu SQ (1985) Research on mineralogy and genesis of pyrite in Jiapigou gold mining district, Jilin Geol 2: 28–35 (in Chinese)

Wyman D, Kerrich R (1989) Archean shoshonitic lamprophyres associated with Superior Province gold deposits: distribution, tectonic setting, noble metal abundances and significance for gold mineralization. Econ Geol Monogr 6: 651–667

Xia ZP (1986) Cambrian geology of Xuanhuacongli of Hebei province and study of gold ore metallogenic regularities. Zhangjiakou Geol 8: 1–48 (in Chinese)

Xu KQ, Sun N, Wang DZ, Liu CS, Chen KR (1982) Two genetic series of granitic rocks in southeastern China. Acta Petrol Mineral Anal 1: 1–19 (in Chinese)

Yang K (1981) Gold leaching experiments of amphibolite rock in the Qinglong district of eastern Hebei province. Gold 6: 28–30 (in Chinese)

Yang LS (1989) The metallogenetic model of Jinchangyu gold deposit in the Archean greenstone belt, Hebei province, China. In: Goode ADT, Smyth EL, Birch WD, Bosma LI (eds) Abstracts and proceedings bicentennial gold '88 May 1988 Melbourne. Geol Soc Aust Abst Ser 23, pp 137–139

Yang LS (1989) Endogenic gold metallogenesis in connection with deep-seated source and prospecting prediction. In: Guan GY, Zhu FS (eds) Proc Int Symp on Gold geology and exploration. 26–30 June 1989 Shenyang. Northeast University of Technology Publ House, Shenyang, pp 664–668

Yang MZ (1988) Tectonics—granites—hydrothermal gold ore belts and the evolutionary geochemical characteristics in northern China. In: Goode ADT, Smyth EL, Birch WD, Bosma LI (eds) Abstracts and proceedings bicentennial Gold '88 May 1988 Melbourne. Geol Soc Aust Abstr Ser 23, pp 658–659

Yang ZY, Cheng YQ, Wang HZ (1986) The geology of China. Oxford Monographs on Geology and Geophysics No 3. Oxford University Press, Oxford

Yu CT, Jia B (1989) Study on the genesis of major types of gold deposits and its mechanism of formation in eastern Hebei. In: Contributions to the project of regional metallogenetic conditions of main gold deposit types in China 2: eastern Hebei province. Geological Publ House, Beijing, pp 1–48 (in Chinese with English summary)

Yu RL, Li WL, Gu SZ, Li JL, Wang FZ, Zhao WH, Liu S, Zhang HX (1989) Metallogenetic conditions of major gold ore types and ore-searching orientation in Eastern Hebei. In: Contributions to the project of regional metallogenetic conditions of main gold deposit types in China 2: eastern Hebei province. Geological Publ House, Beijing, pp 99–146 (in Chinese with English summary)

Zhai MG, Windley BF (1990) The Archean and Early Proterozoic banded iron formations of North China: their characteristics, geotectonic relations, chemistry and implications for crustal growth. Precambrian Res 48: 267–286

Zhai MG, Yang RY, Lu WJ, Zhou J (1985) Geochemistry and evolution of the Qingyuan Archean granite-greenstone terrain, NE China. Precambrian Res 27: 37–62

Zhai MG, Windley BF, Sills JD (1990) Archean gneisses, amphibolites and banded iron formations from the Anshan area of Liaoning Province, NE China: their geochemistry, metamorphism and petrogenesis. Precambrian Res 46: 195–216

Zhai YS (1984) Outline of ore field tectonics. Ministry of Metallurgical Industry Publ House, Beijing (in Chinese)

Zhang J (1979) Features of the main types of gold deposits in northeast China and their ore potential, part 1. Research Institute, Jilin Geological Prospecting Company, Ministry of Metallurgical Industry, pp 23–30 (in Chinese)

Zhang QS (1987) Banded iron formations in China. In: Appel PWU, LaBerge GL (eds) Precambrian Iron-formations. Theophrastus, Athens, pp 423–448

Zhang QS, Liu LD, Zhu YZ, Yang LS (1984) Geology and metallogeny of the Early Precambrian in China. IGCP Proj 91. Report of the National Working Group of China. People's Publishing House of Jilin, Changchun

Zhang RY, Cong BL (1982) Mineralogy and T-P conditions of crystallization of Early Archean granulites from Qianxi county, NE China. Sci Sin Ser B 25: 96–112

Zhang WG (1986) Geological feature and genesis of the Gengzhuang gold-silver deposit, Shanxi Province. Shanxi Metallurgical Geol 2: 1–31 (in Chinese)

Zhang ZM, Liou JG, Coleman RP (1984) An outline of plate tectonics of China. Geol Soc Am Bull 95: 295–312

Zhao YZ (1989) Mathematical and mechanical model and prediction of metallogenic structure system and gold concentration in the Niuxinshan region, Kuancheng County of Hebei province. Technical University of Northeast China Publ House, Qinhuangdao (in Chinese)

Zhong FD (1975) K-Ar isochron age of Precambrian rocks in northeast China. Geochemistry 2: 114–122 (in Chinese)

Zhou QH, Fan YX (1989) On ore-forming geochemistry of Jiaojia gold deposit, Shandong province, China. In: Guan GY, Zhu FS (eds) Proc Int Symp on Gold geology and exploration. 26–30 June 1989 Shenyang. Northeast University of Technology Publ House, Shenyang, pp 829–836

Zhou ST (1989) Genesis of gold deposits in Archean metamorphic series of North China Platform. In: Guan GY, Zhu FS (eds) Proc Int Symp on Gold geology and exploration. 26–30 June 1989 Shenyang. Northeast University of Technology Publ House, Shenyang, pp 277–283

Zhu BC (1979) Metallogenic law and ore prospecting direction of Heilongjiang gold deposits. Main types and ore prospecting direction of gold deposits in NE China. Jilin Geology Institute, Ministry of Metallurgical Industry (internal report), pp 1–24 (in Chinese)

Zhu BQ, Chen YW (1984) Features of Pb isotopic composition of ores and evolution of continental crust of China. Sci Sin Ser B 27: 635–646

Zhu FS (1985) The gold ore deposits geology and metallogeny of Precambrian metamorphic complex in China. Gold 6: 1–7 (in Chinese)

Zhu FS (1989) Study on genetic types of gold deposits in China and their basic geologic features. In: Guan GY, Zhu FS (eds) Proc Int Symp on Gold geology and exploration. 26–30 June 1989 Shenyang. Northeast University of Technology Publ House, Shenyang, pp 18–31

Zonenshain LP, Kuzmin MI, Kovalenko VI, Saltykovsky AJ (1974) Mesozoic structural-magmatic pattern and metallogeny of the western part of the Pacific belt. Earth Planet Sci Lett 22: 96–109

Subject Index